Advances in Anatomy
Embryology and Cell Biology

Vol. 72

Editors
F. Beck, Leicester W. Hild, Galveston
J. van Limborgh, Amsterdam R. Ortmann, Köln
J.E. Pauly, Little Rock T.H. Schiebler, Würzburg

Haide Breucker

Seasonal Spermatogenesis in the Mute Swan (Cygnus olor)

With 30 Figures

Springer-Verlag
Berlin Heidelberg New York 1982

Priv.-Doz. Dr. Haide Breucker
Department of Microscopic Anatomy
University of Hamburg
Martinistraße 52
D-2000 Hamburg 20
FRG

ISBN-13:978-3-540-11326-3   e-ISBN-13:978-3-642-68460-9
DOI: 10.1007/978-3-642-68460-9

Composition: Schreibsatz-Service Weihrauch, Würzburg

2121/3321-543210

Contents

Acknowledgments

The author is much obliged to Dr. Schwarz, Landesforstmeister, Ministerium für Ernährung, Landwirtschaft und Forsten von Schleswig-Holstein, for a shooting licence; to Mr. P von Schiller, Buckhagen, for his kind help in providing a dozen wild swans; and to Mr. H. Nieß, Baubehörde Hamburg, for regularly catching and delivering swans from the waters in and around Hamburg.

Furthermore the very skilful technical assistance of Miss C. Heim and the help of Miss E. Bryden in translating the manuscript are gratefully acknowledged.

Habilitationsschrift presented to and accepted by the Medical Faculty of the University of Hamburg, 3 December 1980.

1 Introduction

Of all the classes in the animal kingdom, birds represent the best known. There are in total about 8600 living species, and the systematic study of this class is more or less complete. Extensive observations – to a large extent by amateur ornithologists – with respect to geographical distribution, life cycles, demands on and adaptations to the environment, breeding habits, migration, and so forth have contributed towards basic and more widely relevant knowledge, e.g., in the areas of ethology, ecology, and evolution and also in social biology (Hilprecht 1970; Farner and King 1971). Together, all these aspects are affected by the reproductive biology of birds, and studies have therefore been carried out for many years with special emphasis on this subject. However, until now this emphasis in avian reproductive biology has been physiological and in particular endocrinological (Murton and Westwood 1977; Roosen-Runge 1977). The morphology of the gonads has been treated in far less detail, and has been confined to a comparatively small number of species, compared with other classes of vertebrates.

Reproduction is the section in the life cycle of an animal which is most dependent upon environmental conditions. Reproduction therefore usually takes place at a particular time, when stress for the adult animals is at its lowest and the chances of survival for the newborn are at their highest, i.e., when light and temperature are optimum for the particular species and a plentiful supply of food is available. Many cold-blooded animals can adapt themselves to such periods, interrupting reproduction during times of unfavorable life conditions either by employing a rest period or by encystment. There are even some mammalian species with similar ability to delay reproduction, either by means of retarded implantation or by means of a latent phase during the embryonal period (Enders 1963; Sadleir 1978). Since in most environments seasonal changes take place in a particular sequence, however, seasonal limitations to reproduction can occur as a special form of adaptation. This form of adaptation is actually very widespread in the animal kingdom. Among birds, a predominant number of species have adapted to changing environmental conditions by means of seasonal breeding periods, because once the development of the embryo has begun interruption is not possible. The early origin, fundamental nature, and constant recurrence of this phenomenon are demonstrated by the fact that there are avian species, e.g., several sea birds and tropical rain forest birds, that still follow the principle of seasonal breeding even when there is no necessity for it because they live in comparatively uniform environmental conditions the whole year round.

In seasonal breeders reproductive activity in both the male and female is restricted by the fact that mature gametes are produced only during a limited period of time. In the case of the male bird this means that the completion of spermatogenesis occurs within a few weeks or months. After this phase, a regression of the testis to a condition of quasi prepuberty takes place, from which a new development of gametes proceeds before the beginning of a new breeding period. This change between the production of gametes and the involution of the gonads is frequently extended over a whole year, and is known as the annual cycle of reproduction. During the regression period the seminiferous epithelium is reduced to a stem cell population of spermatogonia, since all maturation and differentiation processes stop and all the other cells at advanced stages of development are broken down and removed.

The swan *Cygnus olor* Gmelin, a wild or scarcely domesticated species, is a notable member of the seasonal breeders. The light and electron microscopic examinations of the male gonads presented here deal mostly with the seminiferous tubules and touch upon the interstitial tissue only at the periphery. They have been undertaken with the goal of (1) describing fully the course of spermatogenesis, including the differentiation of spermatids, as hitherto in birds this has hardly been dealt with; (2) describing the changes in the seminiferous tubules in the course of a year; and (3) in particular analyzing the process of involution. The seasonal activity in the testis demonstrates how a production of spermatozoa can occur within a limited time through the building up and breaking down of the seminiferous epithelium, without damaging the mechanism responsible for it. This building up and breaking down also occurs in nonseasonal breeders, but is renowned for being obscured by the continual proliferation of gametes. It is, however, important to know about these processes in order to separate them from pathological phenomena. On a wider scale, the building-up and breaking-down process is a generally valid example of a widespread biological phenomenon in which surplus or defective cells present in numerous developmental processes can be eliminated.

2 Literary Synopsis

2.1 Spermatozoa and Spermatogenesis in Birds

The investigation of animal spermatozoa and of their development began about 1840 when, as a result of the newly founded theory on cells, the "small sperm animal" came to be recognized as a cellular element. Based on Kölliker's fundamental observation on invertebrate animals (1841), similar studies were soon being carried out among all classes of animals, particularly the vertebrates. Schweigger-Seidel (1865) gave a pioneer description of the sperm corpuscle (*Samenkörperchen*) found in amphibians, birds and mammals. He was the first to ascertain that a sperm comprises a whole cell. He contradicted Kölliker's idea, which was that spermatozoa were nuclear structures formed in other cells. Schweigger-Seidel had previously observed that in the case of birds and mammals two sorts of cells were produced in the seminiferous tubules, of which only one type developed further to become spermatozoa. From his studies of the domestic cock and the chaffinch, *Fringilla coelebs*, which confirmed observations made by Leuckart (1853), he was able to state that in the avian class two types of spermatozoa are to be distinguished, either being straight and rod-shaped or, as in the case of the songbirds, having a corkscrew-type spiralled head. A somewhat more exact differentiation of the individual segments of avian spermatozoa was given by von Brunn (1884) in his studies of the house sparrow. Like Schweigger-Seidel, he distinguished two sections in the head, as well as a connecting piece, a main piece and an end piece of the tail. Whereas Schweigger-Seidel thought that the connecting piece, which he called middle piece, was part of the head, von Brunn thought it belonged to the tail and therefore recognized the homologous relationship with mammalian spermatozoa.

Moreover, as a separate topic he also dealt with the development of sperm from round cells, i.e., the differentiating phase of the spermatids.

The first thorough and taxonomically comprehensive work on the subject of avian spermatozoa was produced by Ballowitz (1888), who studied 42 species from seven very different orders. His special interest was the tail as an organ of movement, and he intended this investigation to be "a contribution to the theory of the structure of contractile elements." Above all he was able to demonstrate that the axial fiber within a mantle of protoplasm, already described by von Brunn (1884), consisted of further subunits in the form of elementary fibrils which varied in number from species to species and possibly depended on the size of the spermatozoa. However, in his opinion it was the very existence of the fibrils rather than their number and length that was likely to be of decisive importance for contractility. An essential supplement to the findings of Ballowitz (1888) is represented by the investigations made by Retzius (1909), which he carried out over several decades. He made observations of spermatozoa and their development from nearly all animal classes, including many avian spermatozoa. Because of the thorough description of the tail structure already given by Ballowitz, Retzius concentrated on the peculiarities of the sperm head. Apart from this, special interest since the turn of the century has been directed to the chromosomes in the spermatozoa and their behavior during the processes of mitosis and meiosis. While Guyer (1909a, b) on the basis of inadequate technique still described two types of spermatozoa with a different number of chromosomes both in the guinea-fowl and in the domestic fowl, much more exact descriptions are given for the cock by Miller (1938) and for the house sparrow by Riley (1938).

Special notice should be given to the extensive work of Schöneberg (1913) on the development of the male germ cells in the duck. With reference to Regaud's (1901) account of spermatogenesis in the rat, he described a sequence of stages in the seminiferous epithelium during the active phase in spring, and the differentiation of the spermatids. In spite of some incorrect observations, mostly with respect to the spermatogonia and the Sertoli cells, this work can be taken as a basis for our light microscopical knowledge of the spermatogenesis of a non-passeriform avian species. Schöneberg also studied the annual morphological changes in the testis of the duck, and in the following decades the seasonal changes between active and resting phases in avian spermatogenesis were one of the main topics of investigation. In the course of these latter studies, however, morphology was not seriously taken into account, and seasonal reproduction was studied more significantly with regard to endocrine and environmentally dependent control.

Only through the introduction of the electron microscope were new, detailed morphological examinations made possible. But they were predominantly examinations of mature spermatozoa or the differentiation of the spermatids, and these were often only studied in parts. The ultrastructural discoveries confirm the earlier observations made with the light microscope that in the avian class there are two different types of spermatozoa, namely, those of the song-birds (Passeriformes), which have corkscrew shaped heads and tails accompanied by a similarly corkscrew shaped sheath (Yasuzumi 1956; Furieri 1962; Nicander 1970; Humphreys 1972), and the other type of spermatozoa which have rod-shaped heads and tails without striking accompanying structures. This latter type is characteristic of all other birds (Non-Passeriformes) so far studied, including the domestic cock, one of the preferred objects of study (Grigg and Hodge 1949; Nagano 1959, 1962; McIntosh and Porter 1967; Lake et al. 1968; Nicander 1970;

Humphreys 1972; Tingari 1973; Okamura and Nishiyama 1976; Gunawardana and Scott 1977; Maretta 1977), and also the duck, pigeon, and budgerigar, along with some of their relatives (Yasuzumi and Sugioka 1971; Mattei et al. 1972; Humphreys 1975a; Maretta 1975a, b; Yasuzumi and Yamaguchi 1977).

The complete process of avian spermatogenesis from stem cell to mature spermatozoon during the seasonal reproductive period has until now only been dealt with in a few light microscopic works: in the duck (Clermont 1958), the quail (Yamamoto et al. 1967), and the cock (de Reviers 1971). However, in these works the most important point of discussion was the question of the renewal of the stem cells and the morphological basis of a possible kinetics of the germinal epithelium. Experience with mammals, especially laboratory rodents (Leblond and Clermont 1952b), has encouraged this direction of study. Stimulated by Clermont's results (1958) on the germ cell development of the duck, Marchand (1977) has been the only one to describe the ultrastructure of the complete process of spermatogenesis, and he too bases his work on a domesticated species of duck.

2.2 The Annual Cycle of Male Germ Cell Development

Even Aristotle (see Aristoteles 1868) noticed that, especially in the case of birds, an outstanding criterion of the seasonal changes in the reproductive process of male animals was the enormous change in the size of the testes. Even in the twentieth century, the morphological descriptions of annual cycles of spermatogenesis were often confined to those externally visible changes which were repeatedly quantified by weight, volume or length (Disselhorst 1908; Wright and Wright 1944; Hiatt and Fisher 1947; Berthold et al. 1972). Schöneberg (1913) had already stated that, because of considerable individual variation, the increase and the decrease in the size of the testis only gives a rough picture of the process involved. Through his examination of the annual cycle in the duck he came to the conclusion that instead of changes in size the histological changes in the germinal epithelium are much more important. Similarly, in the following decades, the building up and breaking down of the seminiferous epithelium was examined in several avian species (Bissonnette and Chapnick 1930; Kirschbaum and Ringoen 1935/1936; Blanchard 1941; Bullough 1942; Hiatt and Fisher 1947; Blanchard and Erickson 1949; Johnston 1956; Johnson 1961; Lofts et al. 1966a; Payne 1969; Haase 1973; Marchand and Gomot 1973b; Chan and Lofts 1974). However, the process of regression was only given appropriate attention in a very few cases (Bullough 1942; Payne 1969).

The increase and decrease in the diameter of the seminiferous tubules has also been taken into account in the description of changes during the annual cycle (Marshall 1949a; Threadgold 1956/57a, b; Benvenuti 1970). In studies regarding the fundamentals of the annual rhythm such measurements haven often been used as criteria complementary to the valuation of the morphological and hormonal state of the interstitial tissue (Marshall and Coombs 1957; Lofts 1962).

The question which endogenous and exogenous factors are responsible for the change between activity and regression in the male gonads of birds has been the main theme for numerous investigations during the last few decades. A review of the particularly abundant literature on this topic is given by Murton and Westwood (1977). First of all the effect of light on reproduction has been subject of some detailed works

(Lofts and Murton 1968; Farner and Lewis 1971; Menaker 1971), as well as of numerous experimental studies (Bissonnette 1930; Benoit 1936; Lofts and Coombs 1965; Hamner 1966; Lofts 1970; Murton et al. 1970; Schwab 1971; Farner and Lewis 1973; Gwinner 1973; Murton and Kear 1973; Rutledge and Schwab 1974). Humphreys (1975b) is the only one who has followed the ultrastructural changes in the testis of the budgerigar by using an experimental photo-induced cycle. Besides light, temperature (Burger 1948; Engels and Jenner 1956), and rain play an important role, as observations of birds from extremely dry regions show (Keast and Marshall 1954; Serventy 1971). A synergism of the different environmental factors is especially dealt with in several works (Marshall 1949b, 1961b, 1970; Lofts and Murton 1966). Many authors, however, have emphasized that these factors presumably only affect the timing of an autonomous endogen rhythm (Marshall and Serventy 1958, 1959; Lofts 1964; Immelmann 1971; Menaker 1971; Berthold et al. 1972; Gwinner 1973), and this is confirmed through a series of experiments performed on birds from equatorial regions (Miller 1955, 1959; Marshall and Roberts 1959; Marshall and Serventy 1959; Lofts 1964).

In birds the change during reproduction between the active phase and the rest phase is very marked. Similar processes play a role in other animal classes. This is shown through a number of studies on other vertebrates (Aschoff 1955; Lofts et al. 1966b), especially mammals (Tiba et al. 1968; Glover 1973; Lloyd and Englund 1973; Skinner et al. 1973).

3 Material and Methods

Among the five species of the genus swan (Cygnus), the mute swan *Cygnus olor* Gmelin (1789) is the best known in northern European latitudes. Together with the geese the swans make up a subfamily (Anserinae) which, with the ducks and their relatives are again classified as the family of Anatidae. This family, together with that of the game-birds, makes up the order of Anseriformes (Grzimek et al. 1968).

The mute swan, *Cygnus olor* Gmelin, favors areas with a temperate climate. It is not only an ornamental bird seen on lakes and ponds but is also a wild bird without marked migration tendencies and is widespread in north Germany (Bauer and Glutz von Blotzheim 1968; Hilprecht 1970). During the last ten years, the number of swans in this region increased to an unusual extent, so much that it was necessary at one time to regulate their number by controlled shooting. The ornamental or park swan differs from the wild swan only in that it is more trusting of human beings and can therefore be protected from extreme weather conditions, e.g., by keeping the waters free of ice in winter.

Characteristic external marks of the mute swan, *Cygnus olor* Gmelin, common to male and female, in which it differs from the singing swan, *Cygnus cygnus,* are the S-shaped posture of the neck and the shape, color and almost invariably downwards slanting position of the beak. Whilst the lower lip of the beak, a thin rim of the upper beak, the horny peg at the pointed end of the beak, the skin around the nostrils, the cheek-bone areas, and the unfeathered triangle between the beak and the eyes are all black, the upper beak of the fully grown about eighteen months old bird is bright orange in color. The "hump" is also characteristic of the species (hence the German name for the mute swan, "Höckerschwan", "Höcker" being the German word for hump). It is a rounded unfeathered bulge at the roots of the upper beak in front of the forehead, and is predominantly black but becomes gradually redder on the smooth upper side as the bird gets older. It reaches its maximum size at the time of sexual maturity, and in spring it is especially well developed in males which are ready for mating (Bauer and Glutz von Blotzheim 1968; Hilprecht 1970).

In general swans live together in monogamy. As the young birds live under the guardianship of both parents almost until the next breeding season, and the adherence of the adults to a once accepted breeding area is very developed, the mating pairs frequently stay together for many years. After the death or disappearance of one of the partners a new pairing is, however, quite common (Hilprecht 1970). From a distance the adult males are only to be identified with some degree of certainty when they are with their mate, and this is because in comparison with the female they have a somewhat stronger-looking physical appearance. When there is a group of single swans mixed in with the mating pairs it is hardly possible to distinguish between the two sexes. Conclusive assurance of sex and approximate age (young or adult) can only be obtained by the cloacal test (Bauer and Glutz von Blotzheim 1968).

For the examinations presented here fresh testes from 51 sexually mature swans, *Cygnus olor* Gmelin, were prepared. Of these ten were wild swans from the estuary area of the river Schlei in the north of Schleswig-Holstein. The remaining 41 specimens were caught over a period of several years when it was necessary to control their population, on areas of water in and around Hamburg. The time of the preparations was chosen in such a way that for every month material from at least two swans was available. The rest of the birds, including the wild swans, were prepared in the more interesting months of the annual cycle, i.e., in February (three), April (nine), May (seven), June (one), August (one), September (one), October (one), November (three), and December (one).

The wild swans were prepared in the boat on the water immediately after shooting, and the testes were fixed. The park swans were caught live and were then injected with an overdose of Nembutal in the laboratory, beheaded and opened ventrally to reach the testes. Both testes from all 51 birds were examined under the light and electron microscope.

A more exact topographical examination of the abdomen was performed on a further two swans which had died a natural death. To show more clearly the abdominal air sacs and the relationship between their position and that of the testes, each swan was injected through the trachea with some formaldehyde added to an aqueous solution of toluidine blue which, after passing the primary bronchi, colored the air sacs. The birds were opened ventrally and the testes were exposed after removal of the digestive tract.

Light Microscopy: For the light microscopy the following preparations were made: Semithin sections were made from all specimens prepared for electron microscopy and embedded in Epon, and were then colored in an aqueous solution of toluidine blue and pyronine in a proportion of 4:1 with the addition of borax (with reference to Ito and Winchester 1963). Moreover, another set of semithin sections of testes with fully developed seminiferous tubules from April and May were subjected to the PAS reaction. In addition, in April, May, and December halves of left and right testes were fixed in Bouin's solution and then embedded in paraffin in the conventional way. The sections thus prepared were either colored with hematoxyline-eosin or subjected to the PAS reaction (Romeis 1968).

Electron Microscopy: For electron microscopy the material was fixed by immersion, because experience shows this method to be a quicker and safer way than perfusion-fixing and it guarantees a more even soaking of the tissues of the testis. To achieve this, both testes of each bird were first separated from the abdominal air sac, the vas deferens and the blood-vessels, and then sections about 2 mm thick were taken and laid in a 5.5% phosphate buffered solution of glutaraldehyde. A satisfactory hardening of the material takes place during the first quarter of an hour so that small cubes with a maximum side of about 2 mm can be cut from the edges and the center of the sections. These cubes were, at the latest after half an hour, postfixed for 2 hours in 1% phosphate buffered OsO_4 (additionally containing 0.1 M sucrose) and after dehydration embedded in Epon. Semithin sections were processed for light microscopy; ultrathin sections were contrasted using aqueous uranyl acetate and lead citrate (Reynolds 1963). In order to demonstrate the ribonucleoproteins in the nuclei of germ and Sertoli cells some preparations without postfixation in OsO_4 were, according to Bernhard (1968), treated with ethylenediaminetetraacetic acid (EDTA) during contrasting of the ultrathin sections. The electron microscopic examinations were made with a Siemens Elmiskop I, a Zeiss EM 9, and more predominantly with a Philips EM 300.

4 Topography and Microscopic Anatomy of the Testes of the Swan

In birds, unlike most mammals, the testes remain in the same position in the abdomen throughout life. In the adult swan the paired organs are located caudal to the lungs (Fig. 1), separated by parts of the diagonal and horizontal septum; they are situated in the cranial area of the cavitas peritonealis intestinalis. The testes are each attached medially to the back abdominal wall by a short mesorchium. They lie on either side of the dorsal mesenterium, slightly displaced in a lateral direction, ventrally to the caudal end of the adrenal glands and the cranial pole of the heavily lobed kidneys. The testes are further separated from these two retroperitoneal-lying organs by the walls of the abdominal air sacs. These walls, also covered by peritoneal epithelium, are fused dorsally with the peritoneal lining of the back abdominal wall to a far extent laterally. Medially in the region caudal to the testes, this fusion reaches the dorsal mesenterium, but in the area of the testes themselves it only reaches to about the middle of each adrenal gland. Therefore in the swan the testes do not bulge into the

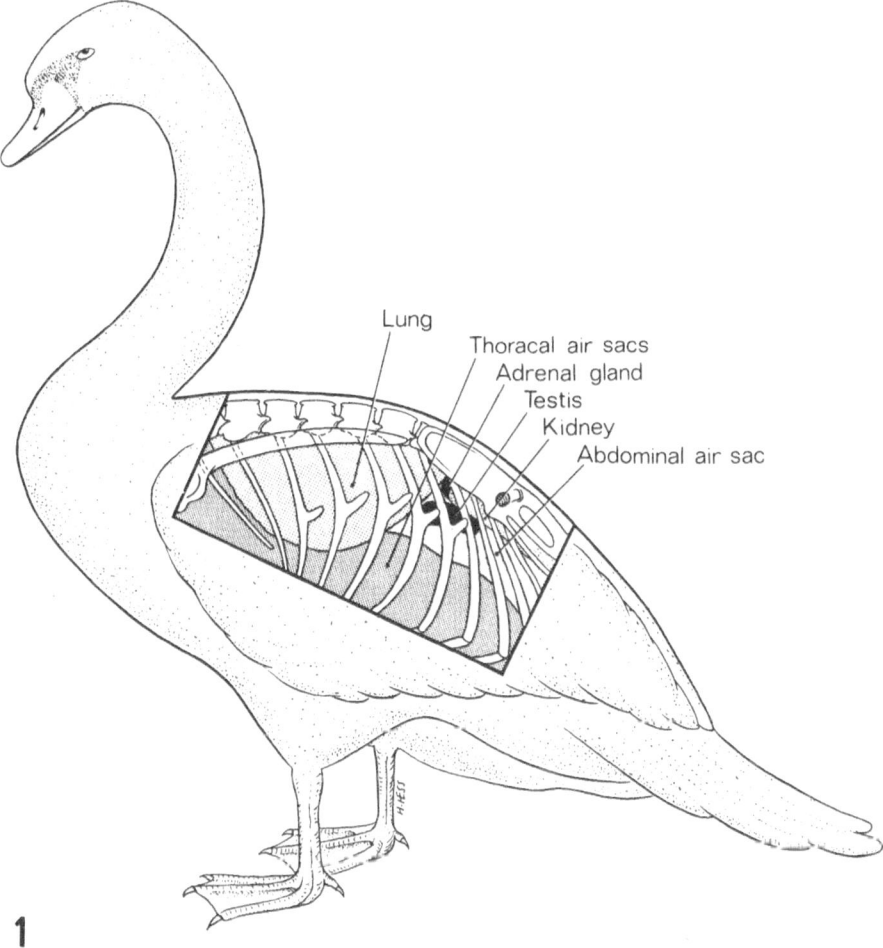

Lung
Thoracal air sacs
Adrenal gland
Testis
Kidney
Abdominal air sac

1

Fig. 1. Topographic relationship of abdominal organs of a male mute swan

abdominal air sacs as they do in other birds (Duncker 1979), but are located on each side in a pocket made from the dorsal mesenterium and the dorsomedial wall of the air sac. This position has only previously been described for the left ovary in the domestic hen (McLelland and King 1970). The wall of the air sac rests displaceably on the surface of the testicle and is unusually delicate in spite of its three layers of air sac epithelium, connective tissue, and peritoneum. However, the assumption of Cowles and Nordstrom (1946) that the abdominal air sacs play a role analogous to that of the scrotum of mammals in cooling the testicle is false. After having passed through the primary bronchi the air is fairly warm by the time it reaches the air sacs (Duncker 1971). Even Lake (1957) agreed with this latter supposition, and Herin and his colleagues (1960), through their experiments on the cock, disproved the theory of Cowles and Nordstrom, although it had already found its way into many articles and handbooks.

The system of excurrent ducts, which is not as well developed in birds as in mammals, undergoes seasonal changes in size similar to those of the gonads themselves (Benoit 1950). Therefore in the swan the epididymis, with its medial and somewhat dorsal position alongside each testis, is macroscopically only visible during the mating period. It is composed of numerous ductuli efferentes, joined to the rete testis, and a short ductus epididymidis. The latter passes with no marked change into the vas deferens, which is ventrally embedded in the kidney over a large area. In the breeding season the vas deferens gets noticeably thicker and longer and seems to be raised from the kidney. It can then easily be recognized in its tortuous course, passing over into the so-called vesicula seminalis and finally meeting the cloaca. Unusually among most avian orders, in the Anseriformes and therefore in the swan a penis is formed; this has the shape of a corkscrew-like pipe and is situated on the ventral side of the cloacal wall. In the nonerect stage it stays concealed in the cloaca.

Because of the asymmetry of the gonads, established already during the embryonal period (Witschi 1935; Dubois and Cuminge 1979), the left testicle in birds is generally bigger than the right one. This is also true in the case of the swan. In spite of the difference in size between the testes, they are alike in shape and the structure of the germinal epithelium and interstitium is the same in each. Both undergo in a similar way the marked seasonal building-up and breaking-down processes. During the extensive regression period in winter they become almost worm-shaped and have an average weight of 150–250 mg and a length of 1–2 cm. In spring during the time of maximum activity their weight increases 40–50 times, i.e., to an average of 7–10 g, and their length increases to 4–5 cm. They then take on a bean-shaped form, of which the concave side corresponding to the hilum lies towards the midline.

Each testis is covered on the outside by a comparatively thin tunica albuginea. In spring particularly this is so delicate that the testis appears whitish-brown in color because the seminiferous tubules show through; in the winter resting period the testis appears somewhat darker, however, pigment cells are only found singly, and then very seldom. Next to the parenchyma of the testis, the collagenous connective tissue fibers of the tunica albuginea are arranged in a thin, straight layer circular to the longitudinal axis of the organ. Towards the periphery the course of the fibers becomes more and more irregular because of numerous blood-vessels. As the tunica albuginea must follow the enormous changes in size of the testis which take place within a few weeks in spring, it may be surmised that in the swan as in other birds (Marshall and Serventy 1957; Lofts et al. 1966a) a renewal or part-renewal of the tunica albuginea occurs each year. This process, however, has not so far been fully attested because the faintly recognizable double-layered capsule, described by the above authors, e.g., in the pigeon, as new and old part of the tunica, in the swan has been observed in all specimens throughout the whole year.

The parenchyma proper of the testis is composed of seminiferous tubules in which the germ cell development takes place. However, the tubules in the swan are not hollow for the most part of the year. Contrary to mammals, in birds this parenchyma is not separated into lobules by septa of connective tissue originating from the tunica albuginea (Lofts and Murton 1973). Because of this the testis can more easily be deformed. In the swan, the connective tissue is limited to few fiber bundles, mostly in the neighborhood of blood-vessels in the angular interstices among the seminiferous tubules as well as in the lamina propria. The absence of separating septa leads to a high incidence of anastomosis between the tortuous seminiferous tubules. The parenchyma of the testis represents an expansive network, so that in the preparation of single tubules only very short pieces can be obtained due to the numerous branchings of the tubules. This is also true for other birds (Lake 1957; King 1975).

The seminiferous tubules, containing the germ cells and the Sertoli cells, are outwardly covered by a basal membrane and beyond this by a mostly very thin lamina propria. The interstitium between the tubules contains the Leydig cells and nerve fibers closely related to them (Baumgarten and Holstein 1968), as well as some scanty connective tissue and vessels.

The complete process of spermatogenesis comprises the germ cell development from the spermatogonia to the free-moving spermatozoa in the lumina of the seminiferous tubules. This phase is reached at the earliest in the 3rd year of life in the swan and thereafter only once a year during a few weeks in spring. From ornithological field examinations (Bauer and Glutz von Blotzheim 1968; Hilprecht 1970) we know that most male swans in any case do not successfully achieve copulation until their 4th or 5th year of life. In the studies reported here it was not possible to ascertain the age of the birds apart from their sexual maturity. Swans live relatively long and therefore may be able to reproduce for at least 10–15 years.

5 The Complete Process of Spermatogenesis

During the period of maximum activity in spring, when the process of spermatogenesis is completed, the organization of maturing germ cells is comparable to that in other birds and also in mammals, i.e., the succession of the individual germ cell generations inside the seminiferous tubules is spatially ordered. The germinal epithelium is made up of four generations of cells which are arranged in order of their progressive development, from the basal membrane to the center of the tubule. Each generation comprises a group of germ cells all at about the same stage of development.

The youngest cells are the spermatogonia lying at the periphery of the seminiferous tubule in contact with the basal membrane. They give rise on the one hand to a pool of stem cells and on the other to the initial material for further development. The following one to three cell layers are made up of primary spermatocytes which, proceeding from an interphase, go into the extremely prolonged prophase of the first maturation division. They go through the leptotene, zygotene, pachytene, and diplotene stages before they enter metaphase. With the completion of the first maturation division the primary spermatocytes turn into secondary spermatocytes. They quickly pass through the second maturation division and give rise to the spermatids. These are the cells found nearest the center of the tubule, which then undergo a long differentiation period. When this phase takes place for the first each spring, the seminiferous tubule forms a lumen in which the spermatids are released at the end of differentiation, as morphologically mature spermatozoa. The seminiferous tubules contain a further cell element besides the germ cells, namely the Sertoli cells.

5.1 Stages of Spermatogenesis

The succession of germ cell generations is determined not only by space but by time as well. Consequently the single steps in the development of the germ cells only occur in well defined combinations; this means that, e.g., a particular stage in the prophase of primary spermatocytes is always and only associated with spermatids at the same definite phase of differentiation. These cell combinations are constant for each species and are known as "stages of spermatogenesis". With a fair amount of certainty at least eight different stages can be recognized in the mute swan (see Fig. 2). Due to the continuous development of all germ cells these cell associations appear at any point of the seminiferous tubules one after the other, and can be placed in a chronological sequence.

The identification of these stages, however, is made more difficult in the case of the swan, in that only groups of few germ cells develop simultaneously, and therefore the single stages generally occupy only small areas of tubular epithelium. Because of this not only do the sections of the seminiferous tubules show widely differing pictures in light microscopic observations, but also in one transverse section of a single tubule more than one stage can be seen juxtaposed. Due to the irregular shape of the stages the cells at the borders of the areas intermingle and give numerous mixed zones with atypical cell combinations which do not fit into any particular stage.

As an indicator for the stages of spermatogenesis other authors have used the single steps of spermatid differentiation. In mammals, the PAS reaction, as a method of representing the acrosome differentiation, was introduced by Leblond and Clermont (1952b) for the classification of the stages. In birds, however, satisfactory results have not been achieved by this method, either in paraffin or in semithin sections. Therefore the stages in the swan, the description of which will follow, were determined in semithin sections principally by examining the nuclear morphology of the spermatids as well as of the primary spermatocytes. The enumeration of the stages was chosen to follow the process of spermatid differentiation.

Stage I (Fig. 2) thus begins with the spermatids which first appear after the second maturation division. They are smaller than the short-lived and therefore relatively seldom found secondary spermatocytes. They have a round nucleus with loosely distributed chromatin in a light cytoplasm. Above these cells in the direction of the lumen of the tubule are much more differentiated spermatids with a heavily condensed, elongated nucleus and an elongated cytoplasm. They are sparsely gathered in bundles and their sometimes recognizable flagella are directed towards the lumen, whereas their acrosomes are directed towards the tubular base.

The primary spermatocytes which are associated with both these generations of spermatids are predominantly in the leptotene stage of meiotic prophase. The chromo-

10

somes are visible as thread-like structures. In the cytoplasm of these cells all cell organelles are positioned in a wide band on one side of the nucleus. The spermatogonia form a one-cell layer along the basal membrane. Their nuclei are round or kidney-shaped. Each nucleus contains a clearly definable nucleolus, and in addition a body

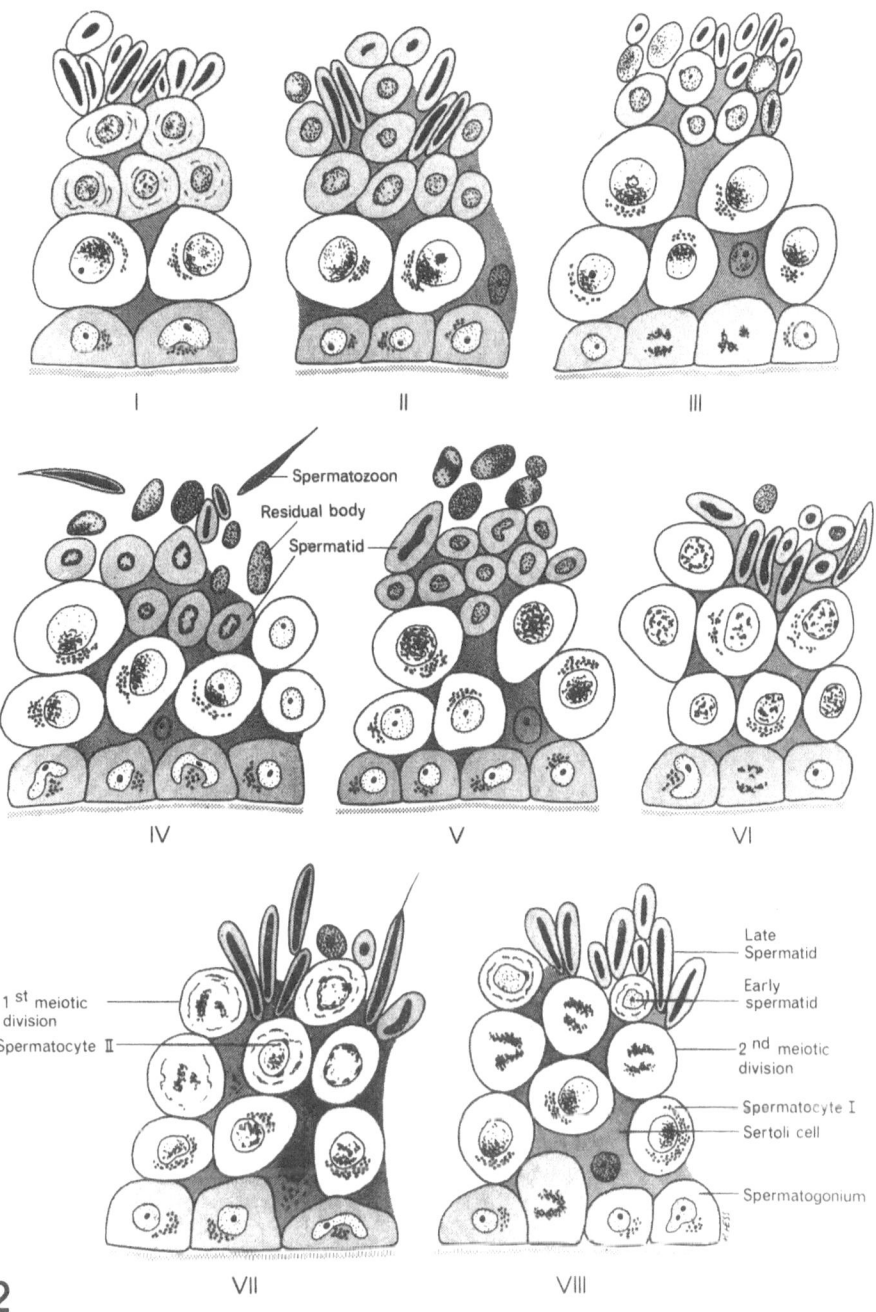

2

Fig. 2. Diagrammatic representation of the eight stages of a spermatogenic cycle in the swan, recognizable at the height of spermatogenic activity on the basis of semithin sections

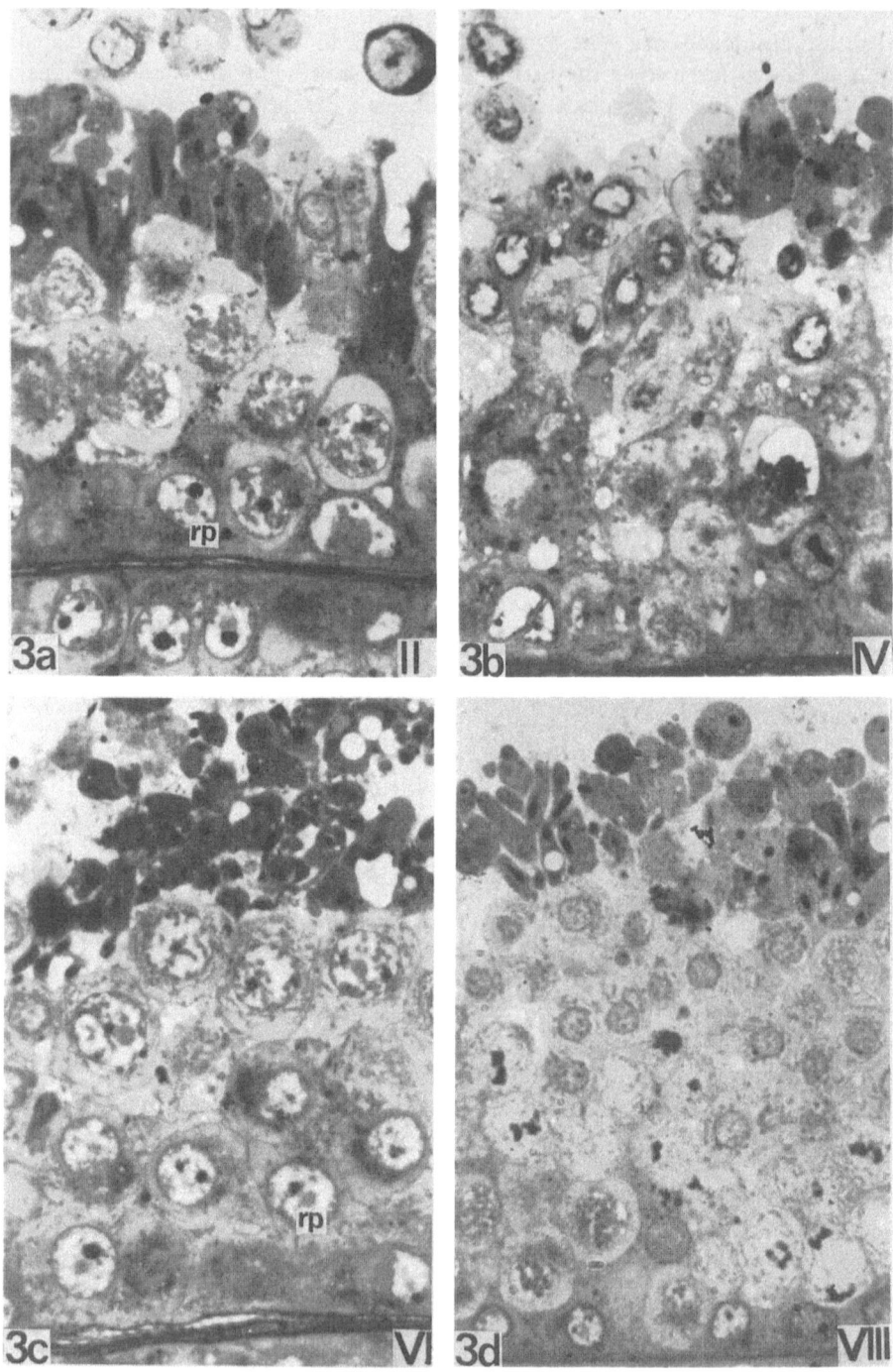

Fig. 3a–d. Light microscopic representation of stages II, IV, VI, and VIII in semithin sections. Sections from seminiferous tubules of a swan at the end of April. *rp*, ribonucleoprotein complex. × 1000

somewhat lighter in color and of nearly the same size (Fig. 3a, c). This is a ribonucleo-protein complex, as the electron microscopic examinations have demonstrated (see p. 20). In the cytoplasm there is a collection of strongly colored particles to be seen, similar to that in the primary spermatocytes, and this is positioned on one side of the nucleus — in the case of the kidney-shaped nucleus in its indentation. Together these particles represent the cell organelles as mitochondria, Golgi apparatus and centrioles. The cytoplasm of the spermatogonia is denser than that of the primary spermatocytes.

In stage II (Figs. 2, 3a) the gathering of the mature spermatids is more noticeable. In section each bundle comprises six to ten spermatids. They are positioned with their pointed end, the acrosome, towards the basal membrane and they are buried in the germinal epithelium between the cells of the generation lying underneath. In the round nuclei of the young spermatids the chromatin begins to position itself along the nuclear membrane, while the rest of the nucleus appears rather light and homogeneous in color. Sometimes one or more dark granules are adjacent to the nuclear membrane on the outside. The PAS reaction does not give any conclusive information as to, whether this relates to the initial stages of the formation of the acrosome. The primary spermato-cytes have, in the meantime, mostly reached the zygotene stage. The chromosomes while entering synapsis move to one side of the nucleus, where the cell organelles have also concentrated themselves in the cytoplasm as an intensively colored complex. The spermatogonia are not noticeably changed. The round nuclei of the cells predominate, however.

Stage III (Fig. 2) is characterized by a loosening of the bundles of the almost mature spermatids shortly before their release from the germinal epithelium. The young spermatids have hardly altered. Their nuclei are still round, somewhat smaller in diameter and with a more distinctly marked nuclear membrane. A tiny granule position-ed on the outside of the nuclear membrane becomes clearer. While this granule is strongly colored with toluidine blue, it is hardly recognizable after the PAS reaction.

In the primary spermatocytes which are still in the zygotene stage, the chromatin has formed a dense aggregation on one side of the nucleus from which single threads can be picked out. The strongly colored complex of the cell organelles still lies in the cytoplasm on the side of the nucleus where the chromatin mass is found. In their basal position the spermatogonia undergo mitosis in this stage, and this sometimes also spreads to the second cell layer. Their equatorial plane is parallel to the basis of the tubule.

In stage IV (Figs. 2, 3b) the mature spermatids are already partly free of the ger-minal epithelium. In addition to spermatids shortly before their release, one finds sec-tions of residual bodies of already freely moving spermatozoa either on the superficial layer of the germinal epithelium or as single bodies in the lumen. The young spermatids begin to stretch. The chromatin of the oval-shaped nuclei is mainly found along the nuclear membrane, but the rest of the nuclei also becomes more dense. At this stage two generations of primary spermatocytes are present. The older primary spermato-cytes are all undergoing zygotene. The chromatin is so densely balled together on one side of the large round nucleus that hardly any single threads can be picked out. The young primary spermatocytes, which have arisen from the mitoses of the spermato-gonia in stage III, are in the interphase. Their nuclei are somewhat bigger than those of the spermatogonia and each contain a clearly definable nucleolus as well as the similarly sized homogeneous and lighter body already mentioned. The latter are also present in the round or kidney-shaped nuclei of the once again inconspicuous spermatogonia. Their cytoplasm is denser than that of the younger spermatocyte generation.

In stage V (Fig. 2) the germinal epithelium on the side of the lumen is covered with differently colored residual bodies, some of which have vacuoles. The following spermatid generation is marked by longish nuclei of irregular shape. The chromatin is distributed over the whole nucleus and is somewhat condensed. Occasionally one end of the nucleus has a stronger color. The cytoplasm has also become more dense and clearly contrasts with the underlying cell layers. The generation of the older primary spermatocytes undergoes pachytene. Between the thickened, loosely distributed chromosomes, which sometimes appear to be split, the ribonucleoprotein complex is again frequently recognizable as a homogeneous body about the size of a nucleolus, but lighter in color. This light-colored body is likewise present in the nuclei of the young primary spermatocytes, which are going through interphase. They also possess a large nucleolus and loosely distributed chromatin and they have grown in size. The cell organelles can be found in a broad band predominantly in one half of the cell. The spermatogonia take an inconspicuous position along the basal lamina and mostly have round nuclei.

In Stage VI (Figs. 2, 3c) nearly all the residual bodies have disappeared. The young spermatids can be recognized because they are more elongated, especially their nuclei. The strongly colored point of the hardly more condensed nucleus which represents the acrosome can be seen more clearly. The shape of the nucleus is sometimes twisted, because the cells are generally still polygonal. During this stage it can be seen that some of the spermatids with their acrosomes become oriented towards the base of the tubule. The older primary spermatocytes show differing pictures. Some of them are in a stage of advanced pachytene and some are in the diplotene stage. During this last period of prophase in which the homologous chromosomes again break away from each other, the chromatin appears as a few dense and more peripheral threads. The young primary spermatocytes have entered the leptotene stage and the chromatin becomes visible as thread-like structures. In the spermatogonia sporadic mitoses occur from which, however, spermatocytes do not necessarily arise. The rest of the cells contain round or kidney-shaped nuclei.

Stage VII (Fig. 2) is different from the previous stage in that here a clear orientation of the spermatids occurs. Each spermatid directs itself with its strongly colored, relatively small acrosome towards the basal membrane, and the flagellum, which is, however, seldom clearly recognizable, towards the lumen. The cytoplasm has become much denser and has followed the longitudinal stretching of the nucleus, which can now extend to its final length. The spermatids gather themselves in small bundles and advance towards the deeper layers of the germinal epithelium.

The spermatocytic population underneath presents a very varied picture. Numerous figures of first meiotic metaphases in the older primary spermatocytes whose equatorial plane is more or less at right angles to the base of the tubule, some cells which are still going through diakinesis, and groups of secondary spermatocytes all appear in this population. The secondary spermatocytes possess a round, somewhat smaller nucleus which contains a loose, more peripheral chromatin, sometimes a nucleolus, and once again the characteristic light-colored body. There are one or more densely colored granules lying close to the nuclear membrane on the outside and these may be related to the chromatoid body. In the cytoplasm the typical concentric garlands of the endoplasmic reticulum which surround the nucleus appear especially clearly. The young primary spermatocytes undergo leptotene. The spermatogonia are without remarkable features.

In stage VIII (Figs. 2, 3d), the last one of the cycle before the beginning of a new sequence, some marked changes are once again to be observed. The spermatids are characterized above all by an advanced nuclear condensation. Thus the nuclei become much thinner. The bundles of spermatids are no longer so distinctive and they are no longer so deeply embedded in the germinal epithelium.

Beneath these spermatids something particularly typical of this stage occurs, namely the division of the secondary spermatocytes, from which is produced a new spermatid generation such as the one found in stage I. The primary spermatocytes are still undergoing leptotene. In the case of the spermatogonia single divisions are also to be seen from which primary spermatocytes do not necessarily arise.

5.2 Incomplete Formation of Stages

The determination of the stages defined is made particularly difficult because of their irregular distribution inside the seminiferous tubules. The latter is the consequence of the extremely variable sizes of the same stages at different sites and the numerous, differently expanded mixed zones with atypical cell associations (Figs. 3a—d). Moreover, the identification of the stages is generally only possible if the tubule has developed a lumen. This is the case only for a very short while, however, and is by no means simultaneous in the whole testicle. Therefore one frequently finds sections of tubules in which the stages are not complete and in which perhaps only three of the cell generations exist. This picture of the incomplete formation of stages is also observed in some tubules with a clearly definable lumen. This suggests that possibly not all cell generations which are present in a complete stage finally develop into mature spermatozoa.

The regular occurrence of definite stages speaks in favor of a controlled procedure of germ cell development, i.e., a continuing process, in this case comprising eight successive cell associations, which can be exactly fixed in time. The time-span in which such a sequence takes place before a new succession of stages begins is known as the "cycle of the germinal epithelium". With the beginning of each new cycle, the single cell generations become displaced to a level nearer to the lumen. Therefore it is assumed that the development from a spermatogonium at the basal lamina up to a spermatozoon which is released into the lumen takes a period of time extending over several cycles (Fig. 2). The question of the amount of time required for a cycle to take place has so far not been settled in the case of the swan.

5.3 Stem Cell Renewal

The open questions regarding the kinetics of spermatogenesis in the swan are for the most part grounded in inadequate knowledge about the renewal of the stem cells in the spermatogonia. In all eight stages the cells along the basal lamina resemble each other so closely that no morphological criterion for the stage of their development can be given. One cannot therefore differentiate between stem cells and those spermatogonia which develop into primary spermatocytes. All cells show the same concentration of cell organelles on one side of the nucleus, regardless of whether the nucleus is round or kidney-shaped. In the same way the nucleus nearly always contains a nucleolus and the homogeneous light-colored body. The different shape of nucleus is possibly just a sign which tells us that the cell is either in interphase or just about to undergo mitosis, because shortly before mitosis takes place round nuclei are predominant. One cannot see from the cells how often they are going to undergo mitotic division before they differentiate. The uniform shape of the spermatogonia also makes an association with the different stages of spermatogenesis impossible. The one fact that can be stated with any degree of certainty is that primary spermatocytes only arise after mitoses in stage III, whereas spermatogonial mitoses in stages VI and VIII, which only take place singly and not in groups, need not necessarily be followed by primary spermatocytes. These mitotic divisions probably therefore serve in the multiplication of stem cells.

5.4 Sertoli Cells

In all stages, besides the generations of germ cells single Sertoli cells are present (Figs. 3b–d). They extend through the full thickness of the germinal epithelium from the basal lamina to the free surface at the lumen of the tubule. Their nucleus is round or oval with no indentation and the distribution of chromatin is homogeneous. Most of them contain a clearly definable nucleolus and frequently the same light-colored homogeneous body as is found in the germ cells. The nucleus is nearly always situated at the level of the second or third layer of germ cells, i.e., in between the youngest developed spermatocytes. In general the cytoplasm of the Sertoli cells does not show any light microscopically visible structures of special importance. Its density is similar to that of the spermatogonia and, with its thin processes, it is therefore well recognizable between the light-colored spermatocytes and the young spermatids. Occasionally the cells in their basal and middle part contain differently colored inclusions which, however, cannot be related to the cycle of the germinal epithelium.

5.5 Discussion

In the swan as an example of a barely domesticated avian species the complete process of spermatogenesis has so far not been described. It could be shown that the development of the male germ cells which takes place in seminiferous tubules follows the rules of spermatogenesis in higher vertebrates. A principle of ordering in the germ cell development, as subsumed under the term kinetics, does exist, but is often hidden because of a great variability within the cell associations, the so-called stages of the germinal epithelium. In this respect a comparison may be allowed to the development of sperm in man, which also does not offer an easy classification of the germ cell development.

If one compares to that of other birds the development from spermatogonia to mature spermatozoa in the swan during the breeding season, the findings are in parts very similar, especially in the case of the duck, on which several detailed descriptions of light microscopic examinations are available (Schöneberg 1913; Clermont 1958; Marchand and Gomot 1973a). But observations on the cock (Miller 1938; Zlotnik 1947; de Reviers 1971) and the Japanese quail (Yamamoto et al. 1967) are also comparable with the findings in the swan.

In addition to the process of spermatogenesis in seminiferous tubules as typical for the amniotes (reptiles, birds, mammals) in general, these species, which all belong to the goose-like or gallinaceous birds, have in common that the development of their germ cells is synchronized only to a limited degree. That means that on the whole only a few cells at the same stage of development lie next to each other. In this they are essentially in contrast to the birds of the finch family, for example, in which one finds germ cells in large areas of the testis at the same stage of development, as already proved by Schöneberg (1913). Similar differences also occur among mammals between primates and rodents (Clermont 1972). Cross sections through seminiferous tubules of the swan reveal in the order of the germ cells a picture more similar to that found, for instance, in man (Clermont 1963) than to that found in the song-birds (Lofts and Murton 1973).

The classification of spermatogenesis into stages, as practised in the fundamental work on laboratory rodents by Leblond and Clermont (1952a, b), can be carried out only with difficulty in the swan. In spite of the irregular distributions and dimensions of the single cell generations, in the goose-like and gallinaceous birds a regular process of development can be recognized in the form of repeated appearances of cell associations. The stages of spermatogenesis in the duck were already given in 1913 by Schöneberg and later by Clermont (1958), and Yamamoto and his colleagues described the stages in the quail in 1967. All authors point out the special difficulty of identifying the stages because of their limited expansion and the consequently frequent appearance of atypical cell combinations in the intermingling zones of two adjacent stages.

The methods of examining the kinetics of spermatogenesis applied to mammals can be used only to a certain extent in birds. Clermont (1958) ascertained the stages in the Peking duck by watching the single steps in the development of the acrosome during the differentiation of the spermatids with the help of the PAS reaction (Leblond and Clermont 1952b), which had proved itself very useful in mammals. Clermont himself, however, noticed that it was not such a good method in the duck. Marchand and Gomot (1973a), using semithin sections, later confirmed Clermont's results without going into details. Cavazos and Melampy (1954), in a comparative study about the applicability of the PAS reaction to the testis of vertebrates, described a number of not very differentiated phases in the development of the acrosome in the cock without relating it to the remaining germ cell development. The weak ability of the acrosome or its precursors to absorb color seems to leave the exact definition of the developmental state of the spermatids in question. Yamamoto and his colleagues (1967) rejected the PAS method for listing the stages in the quail, and de Reviers (1971), from this experience, based his description of the eight stages in the differentiation of spermatids in the cock only on the morphological changes in the nuclei of these cells. A relation to the other cell generations of the germinal epithelium and thereby a definition of the stages of spermatogenesis seemed to him to be impossible, although he presumed their existence. Instead of the inadequate PAS reaction Gunawardana (1977) used semithin sections of material which had been embedded in araldyte, and ascertained ten steps in the differentiation of spermatids in the cock. But he too failed to relate his findings to the remaining germ cell development, and therefore does not get much further than Zlotnik's description of the development of spermatids in the cock (1947).

The two last-named authors describe as a particularly obvious step in the development the temporary multiple contortions of the spermatid nucleus. This feature of the nucleus has also regularly been observed at the beginning of the elongation phase in the swan, as well as in the duck (Clermont 1958; Marchand and Gomot 1973a) (see also electron micrographs Figs. 11a and 12, and Marchand 1977). Therefore the contortions must in this case be one of the normal timed processes in the development of the spermatids and not a degenerative change in the nucleus as de Reviers (1971) suggested.

By means of the valuation of the nuclear morphology in the primary spermatocytes and in the spermatids one can identify eight stages of spermatogenesis in the swan with a fair amount of certainty. The definition of the stages of spermatogenesis, however, cannot be given without a certain degree of subjectivity on the part of each observer, as Clermont (1972) has already stated. The choice of eight stages seems to be correct in comparison with studies done by Clermont (1958) and Yamamoto and his colleagues (1967), who also found eight stages in the duck and the quail respectively. In accordance with Clermont the beginning of each cycle of the germinal epithelium, stage I, has been fixed at the point when a new generation of spermatids just previously arisen out of the second maturation division begins differentiation. Their development into mature spermatozoa and their release into the lumen of the tubule requires more time than the duration of one cycle, but not as long as two complete cycles as in the quail (Yamamoto et al. 1967); for this species, consequently, the beginning of the differentiation of young spermatids coincides with the spermiation of an older generation, which is unusual. Schöneberg (1913) − with reference to one of the first descriptions of a cycle in the rat by Regaud (1901) − based his classification of stages in the duck on the complete differentiation phase of the spermatids, and ended up with 12 stages containing a further set of subdivisions.

The combination of cell generations within the individual stages in the swan agrees mostly with the data of Clermont (1958) on the Peking duck. Slight differences may be due to the use of different methods (Epon against paraffin; morphology of the nuclei of the spermatids instead of the development of the acrosome). Especially striking periods in the development, however, appear in the same sequence, e.g., in stage III mitoses in the spermatogonia, from which a new generation of primary spermatocytes arise; in stage IV the beginning of the spermiation; in stage VIII the end of the second meiotic division.

On the duration of a cycle, i.e., the time required for all the stages to take place one after the other in a given area, there are still no data for any avian species. Obviously, de Reviers (1968) is so far the only person who has gone into the question of the duration of spermatogenesis in birds, without, however, referring to stages of spermatogenesis. By means of autoradiographic tracing he was able to determine the length of time needed for meiotic prophase in the cock, as well as the time up to the release of labeled spermatozoa. Using the same methods he examined the duration of spermatogenesis in the Peking duck in collaboration with Marchand and Gomot (Marchand et al. 1977). In both the cock and the duck the meiotic prophase required 5 1/4 days, and at the end of the 12th day after injecting the radioactive substance into the blood vascular system the first labeled spermatozoa appeared in the ejaculation. This time-span does not, however, comprehend the whole process of spermatogenesis, because on the one hand it includes the duration of the passage of spermatozoa through epididymis and ductus deferens, and on the other the observation of the label only begins with the primary spermatocytes in the leptotene stage. The autoradiographic tracing technique is certainly the method to choose in order to reach precise evidence on the kinetics of spermatogenesis. In the swan it has not been used up until now because of technical reasons (the size of the animal and difficulties in keeping it imprisoned without affecting its natural process of spermatogenesis).

To obtain information on the stem cell renewal in spermatogenesis through morphological findings requires considerable effort. As might be expected from observations on the duck, it is impossible unequivocally to identify different types of spermatogonia in the swan. Clermont (1958) pointed this out in opposition to Schöneberg (1913) and Marchand and Gomot (1973a). De Reviers (1971) described spermatogonia in the cock with round or oval nuclei, but without further classification. Yamamoto and colleagues (1967) also found different forms of nuclei in the quail, and used these to classify three different types of spermatogonia. Their chronological succession and their distribution to the eight stages described in the quail were recorded in a table, but in the text the information was considerably limited.

However, a reliable distinction at least of a stem cell population from the rest of the spermatogonia must be supposed for a quantitative statement about the stem cell renewal. Moreover, only by the numerical association of both the differentiating spermatogonia which give rise to primary spermatocytes, and the stem cells, with the single stages of spermatogenesis is it possible to fix the very beginning of spermatogenesis and thereby to calculate its total duration. The only statement which can be made with respect to the swan is that in stage IV a new generation of primary spermatocytes appears, which have in stage III undergone their last mitosis as spermatogonia before entering into meiotic prophase. However, one cannot discover how long ago these cells became different from stem cells and how often they have divided by mitosis since they became specialized spermatogonia. Clermont (1958), having treated the duck with colchicine and counted the number of mitotic divisions in stages III, V (which corresponds to stage VI in the swan), and VIII, found a ratio 2:1:1. He formed two hypotheses about the stem cell renewal from this discovery, namely that new stem cells were produced either only in stage V or in both stages V and VIII, and that the mitotic divisions in stage III only gave rise to primary spermatocytes. If one presumes on the basis of experience with mammals (Roosen-Runge 1962; Clermont 1972) that in birds a corresponding space of time is required for the development of the specialized spermatogonia into mature spermatozoa, namely somewhat more than four cycles, then the differentiated spermatogonia, which are no longer stem cells, must go through at least two or three mitotic divisions. Comparable conditions may be operative in the swan.

In summary it can be stated from the results in the swan that such a regular ordering of the germ cells in the seminiferous tubules as found, e.g., in the rat and mouse, does not exist in the group of geese and gallinaceous birds. The apparent disorder is similar to that observed in several mammals including man, and makes an analysis of the kinetics of spermatogenesis more difficult. From this fact it is possible to infer that an especially highly developed principle of order in the process of germ cell development is not necessarily essential for the general biological phenomenon of spermatogenesis, although the results of studies on rodents have been decisive for the methods and success in research of kinetics.

The Sertoli cells seen in examinations of avian spermatogenesis made with the light microscope have not received much attention other than the mere mention of their existence. De Reviers (1971) only followed their increase in number during the ontogenesis of the cock. A somewhat more detailed description for the duck was given by Schöneberg (1913). It is true that he still took the total of Sertoli cells for a syncytium but he emphasized the typical position of their nuclei between the nuclei of the spermatocytes, in opposition to findings in the rat, where the nuclei were directed more towards the tubular basis. This fact has also been mentioned by Clermont (1958) and can be confirmed in the swan during active spermatogenesis. All the authors give the same description of irregularly shaped nuclei as characteristic of the Sertoli cells. This fact, however, is absolutely not true in the case of the swan. Here, the nuclei of the Sertoli cells are striking precisely because of their consistently regular shape, which can be observed independently of possibly differing functional states. In relation to the maturation of the spermatids and their temporary bundling, the Sertoli cells are seen as supporting cells (Schöneberg 1913; Zlotnik 1947). In the swan, the light microscopic examinations of the Sertoli cells scarcely show new aspects. These come to light only on the basis of further morphological details discovered by means of electron microscopic observations (see p. 41).

6 The Ultrastructure of the Seminiferous Tubule

During the short phase of sexual activity the germ cell development can be described light microscopically as a succession of defined stages. A detailed representation of the different cell types is only possible, however, in the area of ultrastructural examination. In the following the individual cell generations which make up the complete process of spermatogenesis should be characterized in their fine structure.

6.1 Spermatogonia

The spermatogonia are predominantly found only in a single layer along the base of the seminiferous tubule (Fig. 4). They have a not quite uniform appearance; nevertheless the distinction of different types is just as difficult with the electron microscope as with the light microscope. The cells are first of all recognizable because of their contact with the basal lamina. They do not always rest upon it with a broad base, but often touch it only with more or less broad feet which are separated from one another by Sertoli cell branches (Figs. 5, 6). Sertoli cells also frequently separate spermatogonia from each other laterally, if they are not joined by intercellular bridges. Occasionally such bridges also arise between cells of the basal and the second layer.

The extent of the basal bearing area often depends on the shapes of the cells which are themselves related to the multiplicity of nuclear shapes. The nuclei are round, oval or, frequently, bent, in which case they are either kidney-shaped or elongated, usually having a thinner end and a fatter end. They are positioned lengthways either at right angles or parallel to the basal lamina (Figs. 4, 5, 6). The chromatin is predominantly distributed loosely along the nuclear membrane, so that on the whole the nuclei appear very transparent. In general one can observe one, seldom two, nucleoli, which are made up of two different components, a very electron-dense reticular structure, and a compact area of lower contrast (Figs. 4, 5). Next to them one usually finds a dense aggregation of 22—25 nm granules (Fig. 6); these are ribonucleoproteins, as can be specially demonstrated after treatment with EDTA (Bernhard 1968). Together they correspond to that homogeneous body about the size of a nucleolus, which is recogniz-

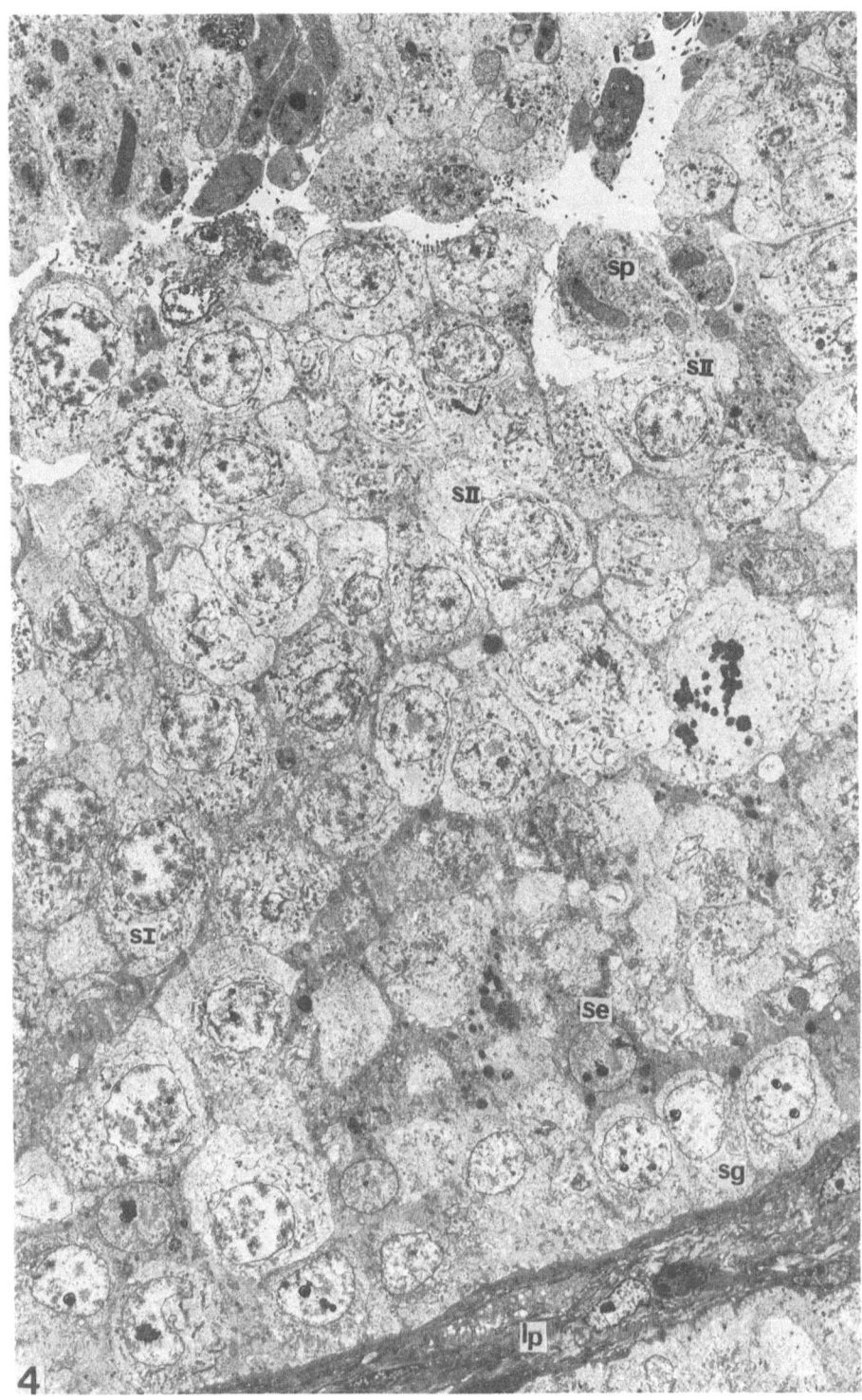

Fig. 4. Section from a seminiferous tubule in April. *lp*, lamina propria; *se*, Sertoli cell; *sg*, spermatogonium; *sI*, primary spermatocyte; *sII*, secondary spermatocyte; *sp*, spermatid. × 1450

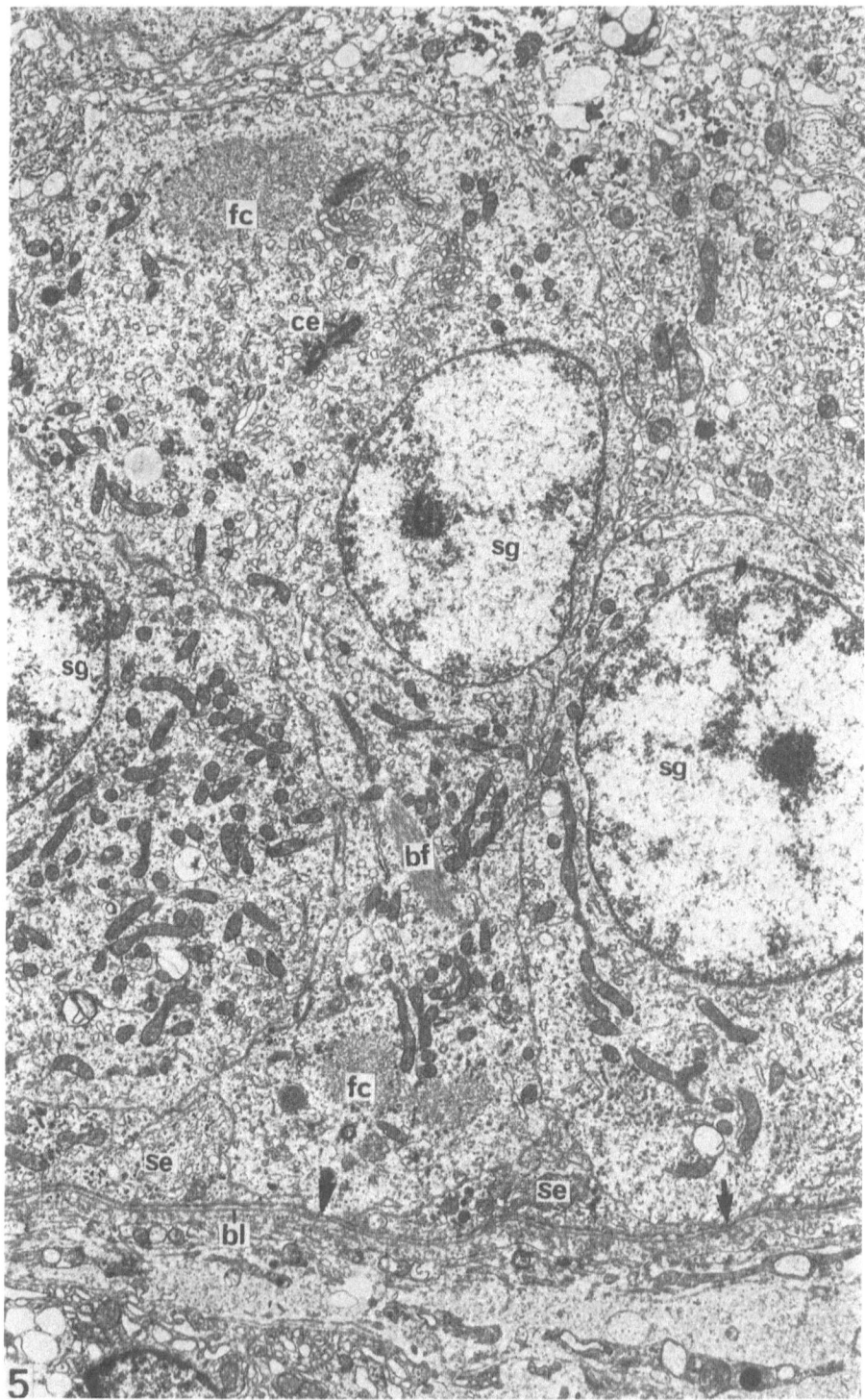

Fig. 5. Basal compartment of a seminiferous tubule at the beginning of May. The spermatogonia show only small areas of contact (*arrows*) with the basal lamina. *bf*, bundles of filaments; *bl*, basal lamina; *ce*, centriole; *fc*, filamentous-granular complex; *se*, Sertoli cell; *sg* spermatogonium. × 7200

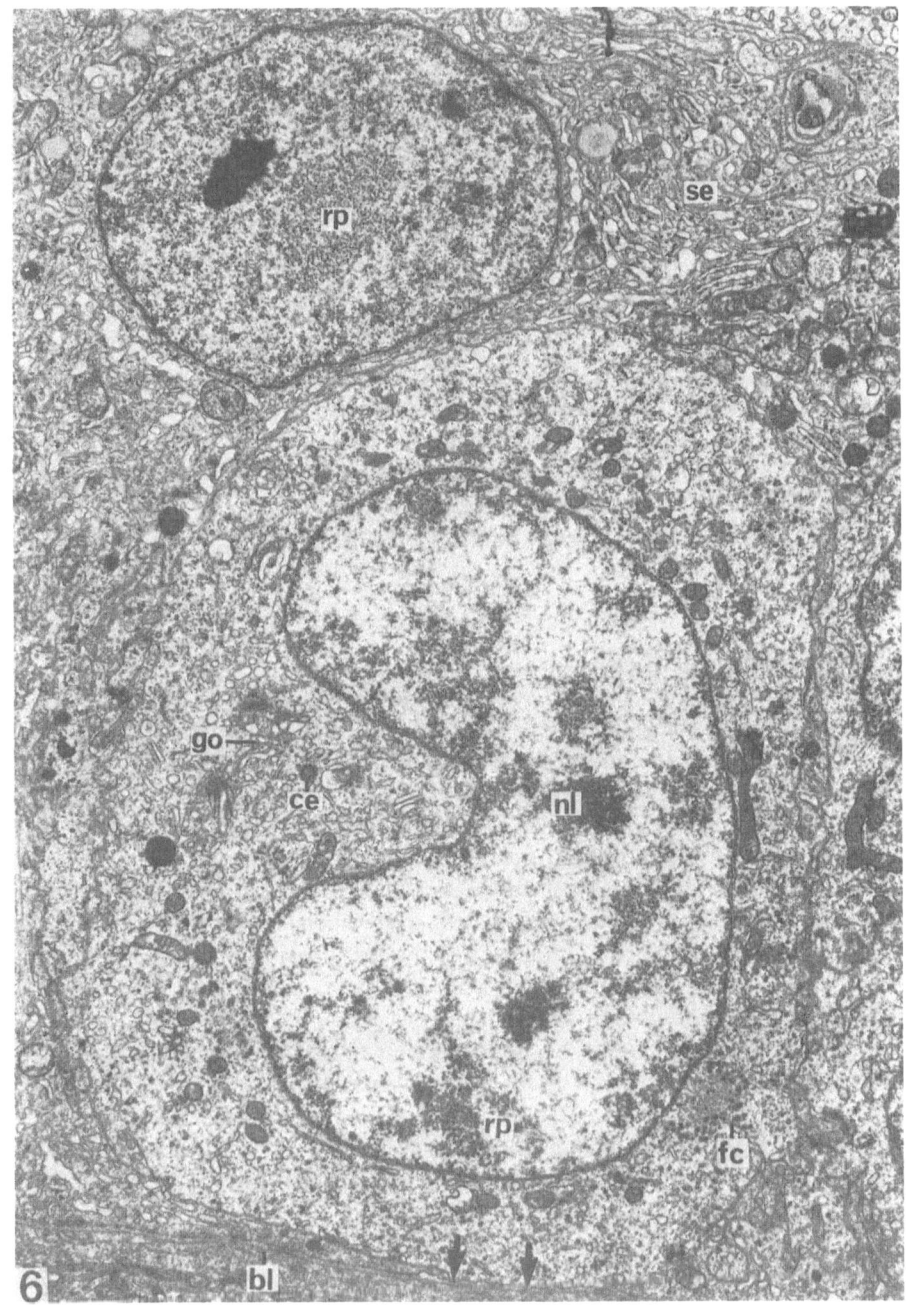

Fig. 6. Basal region of a seminiferous tubule at the beginning of May. There are only points (*arrows*) of contact with the basal lamina. *bl*, basal lamina; *ce*, centriole; *fc*, filamentous-granular complex; *go*, Golgi apparatus; *nl*, nucleolus; *rp*, ribonucleoprotein complex; *se*, Sertoli cell. × 8800

able under the light microscope, but appear lighter than a nucleolus also when seen under the electron microscope. Sometimes there are more than one of these bodies.

The cytoplasm of the spermatogonia is rich in free ribosomes, which are mostly associated as polysomes; the endoplasmic reticulum is found with and without attachment of ribosomes. The small, somewhat elongated mitochondria are predominantly concentrated on one side of the nucleus, and several of them are positioned very closely to one another, but no intermitochondrial cementing material has been found. Amongst the accumulation of mitochondria an extended Golgi apparatus can often be seen, which again may enclose two centrioles standing at right angles to each other (Figs. 5, 6). As a characteristic of the spermatogonia one occasionally finds next to single long microtubules a bundle of very fine filaments in the cytoplasm, each one with a diameter of 6–8 nm (Fig. 5). Moreover, one or several complexes regularly appear, which are made up of a fine granular-filamentous material; this is surrounded peripherally by very electron-dense granules, each 60–70 nm in diameter (Figs. 5, 6). It is possible that this complex is some kind of elementary form of the chromatoid body, or a special form of the "nuage" which is frequently found in early germ cells. All the cell components can be observed in the spermatogonia, either together or individually and independently, without being related to the shape of the nuclei of the cells.

6.2 Primary Spermatocytes

After the spermatogonia follows the generation of primary spermatocytes. They no longer have any contact with the basal lamina (Fig. 4). They grow during interphase, before entering into meiotic prophase, to become the biggest cells of the germinal epithelium with a nuclear diameter of 8–10 μm. They can be joined to each other by intercellular bridges. During leptotene, the chromosomes because of their condensation appear as a dense aggregation of chromatin. With the transition into zygotene the nucleolus disappears, and the chromosomes are positioned on one side of the round nucleus while the rest of the space in the nucleus appears electron-optically almost empty (Fig. 7a). Occasionally arising vacuoles at the nuclear membrane point to the process of condensation. During zygotene the pairing of the homologous chromosomes begins, which is recognizable by the forming of typical synaptonemal complexes. With the further displacement of the chromosomes to one side of the nucleus, the mitochondria, the Golgi apparatus, and the centrioles appear more closely pressed together than in the spermatogonia and are concentrated on one side of the nucleus, namely the side where the chromosomes are found (Fig. 7a). Near the end of zygotene the EDTA-positive granular complex is again clearly visible (Fig. 7a), having been hidden during leptotene and zygotene. In the following pachytene, the chromosomes appear more and more as single elements which are again distributed over the whole of the nucleus. The homologous partners stay together and the synaptonemal complexes become more clearly visible, as does the granular ribonucleoprotein complex (Fig. 7b). In the cytoplasm the aggregation of mitochondria, Golgi apparatus, and centrioles is no longer conspicuous, and the cell organelles partly are arranged already in the way as is characteristic for the secondary spermatocytes. In this case, the smooth endoplasmic reticulum is mostly formed into long chains, which are frequently composed of vesicles fastened together like a string of pearls, and to these single mitochondria are attached (Fig. 7b).

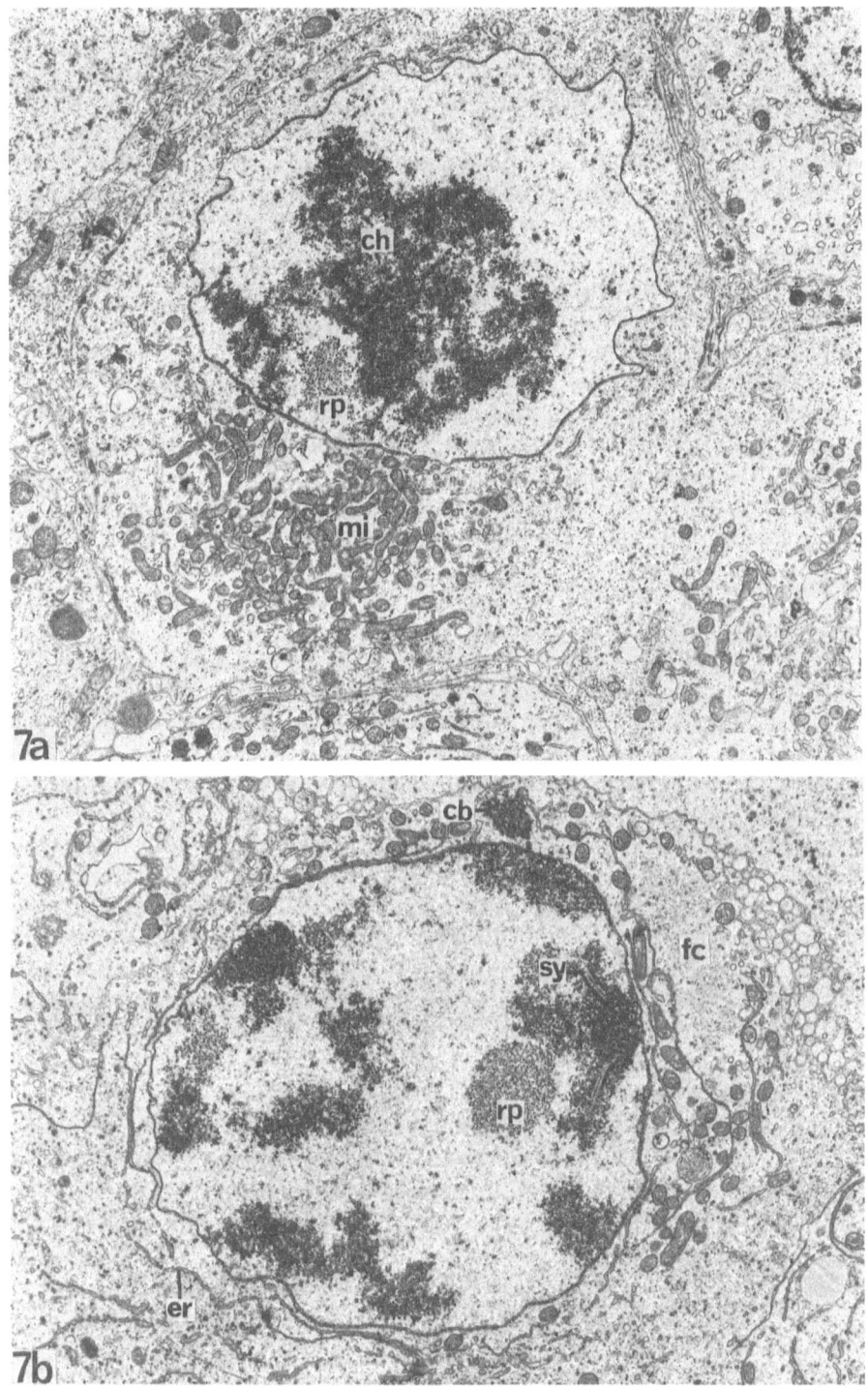

Fig. 7a, b. Primary spermatocytes at the beginning of April; *a* zygotene, *b* pachytene with characteristic arrangement of endoplasmic reticulum. *cb*, chromatoid body; *ch*, chromosomes; *er*, endoplasmic reticulum; *fc*, filamentous-granular complex; *mi*, mitochondria; *rp*, ribonucleoprotein complex; *sy*, synaptonemal complex. *a* × 7000, *b* × 7600

25

Furthermore, the chromatoid bodies – mostly more than one per cell – attract attention in pachytene (Fig. 7b), as they now look similar to the well known picture found in mammals. They consist of contrasting areas of partly fibrous, partly granular material and can mostly be found near the nucleus, often actually in contact with the nuclear membrane which in this area exhibits an especially numerous quantity of nuclear pores. Sometimes the chromatoid body can also be found in contact with the filamentous-granular complex, which has already been seen in the spermatogonia. Whether its peripheral granular component, which in some cases is no longer recognizable, becomes part of the chromatoid body cannot be stated with certainty. The substructure of the chromatoid body seems at this point rougher and more irregular and is therefore not clearly measurable.

The pachytene is followed by the diplotene and diakinesis. As these phases are not very frequently found they probably take place within a comparatively short space of time. They are marked by the disappearance of the synaptonemal complexes and the displacement of the chromatin towards the periphery of the nucleus. The mitochondria, in contact with the smooth endoplasmic reticulum, distribute themselves over the whole cell; the centrioles stay in close proximity to the Golgi apparatus until the beginning of metaphase, and after the division they resume this former position.

6.3 Secondary Spermatocytes

The secondary spermatocytes also appear only briefly. Their round nucleus measures 5–6 μm in diameter and contains few chromatin accumulations, mostly in the peripheral areas. Occasionally a compact nucleolus or one with a lighter-colored center is present, and rather regularly there are one or several EDTA-positive bodies. The mitochondria, associated with the garlands of smooth endoplasmic reticulum, are characterized by their light matrix and densely packed contrasting cristae. A striking feature of these cells are the so-called annulate lamellae (Fig. 8a), which can be seen here in nearly every cell, although they occasionally appear also in earlier germ cell types, even in spermatogonia (Figs. 24, 27b). They are cisternae the double membranes of which are penetrated at very regular intervals by pores or annuli and which are arranged in several parallel layers mostly in contact with the nucleus. Their ends frequently pass over into the cisternae of the smooth endoplasmic reticulum. Between them and the nuclear membrane or between two of these layers themselves a chromatoid body can sometimes be found (Fig. 8a).

Apart from the chromatoid bodies the filamentous-granular complex is often to be found some distance from the nucleus, but mostly the peripheral component is missing. Independently of the one or two centrioles in the region of the Golgi apparatus, single cilia without an outer membrane of their own occasionally appear in these cells, but disappear again during the development of the spermatids.

6.4 The Differentiation of Spermatids

The spermatids which result from division of the secondary spermatocytes are only slightly smaller than the previous cell generation. At the beginning of their differentiation the spermatids possess a round nucleus with a diameter of 4.5–5 μm, the chromatin of

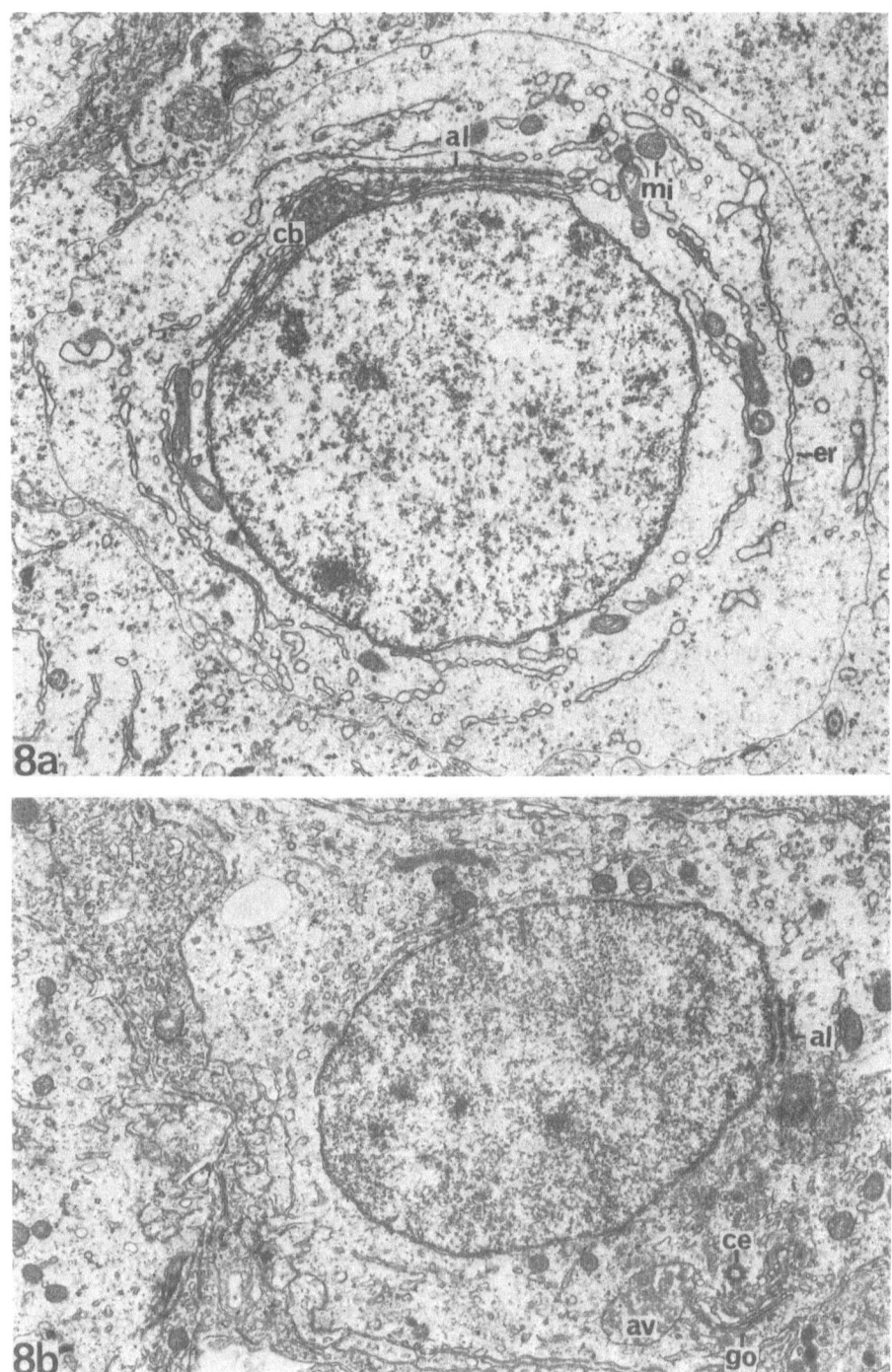

Fig. 8a, b. Secondary spermatocyte *a* and early spermatid *b* from the middle of April. *al*, annulate lamellae; *av*, acrosomal vacuole; *cb*, chromatoid body; *ce*, centriole; *er*, endoplasmic reticulum; *go*, Golgi apparatus; *mi*, mitochondria. *a* × 11 400, *b* × 10 600

which is much more evenly distributed than that of the secondary spermatocytes. As well as several small nucleoli the EDTA-positive body is recognizable. In the cytoplasm, which is frequently joined to neighboring cells by intercellular bridges, the garland-shaped organization of the smooth endoplasmic reticulum and its close contact with the mitochondria is broken up. The endoplasmic reticulum disintegrates into numerous single vesicles. The rest of the cell contents, such as free ribosomes, Golgi apparatus and centrioles, filamentous-granular complex, and annulate lamellae, are all still present as previously.

The first clear sign of the beginning of a differentiation is a vacuole containing a few single grana or a very loosely distributed filamentous material, which can be observed in the cytoplasm. It is the acrosomal vacuole which appears in close contact with the Golgi apparatus (Fig. 8b). At this phase, the vesicles of the latter are mostly filled with homogeneous material of median electron density. In the next, the acrosomal vacuole has achieved contact with the nuclear membrane while the filamentous contents have increased in density (Fig. 9a). During this time the centrioles remain in the region of the Golgi apparatus. At the point where the acrosomal vacuole contacts the nuclear membrane the latter has become indented and strengthened, as on the inside electron-dense material has positioned itself there and the nuclear pores in this area have disappeared (Fig. 9a).

Meanwhile the development of the spermatid tail from the two centrioles has also begun, the latter usually being orientated exactly perpendicular to one another (Fig. 9b). The first steps of differentiation are recognized in the area of the Golgi apparatus, to begin without contact with the nuclear membrane. Instead of this, the future distal centriole, touches the cell membrane (Fig. 9b). This distal centriole grows into a hollow cylinder by the increased length of which the proximal centriole, with a collection of Golgi vesicles, seems to be pushed towards the nucleus. The hollow cylinder shows in cross section the typical configuration for a basal body of nine peripheral triplet microtubules, and contains in its proximal third an electron-dense core, sometimes separated by a lighter cleft. Moreover, this centriole is the beginning of the axoneme which extends outwards from the cell membrane (Fig. 9c). At the transitional zone between the flagellum and the cell membrane a small amount of electron-dense material has aggregated, which represents the annulus (Figs. 9b, c). It is made up only of one component and indicates the end of the future middle piece. The proximal centriole likewise consists of nine peripheral triplets with a light center (Fig. 14a). On its surface directed towards the nucleus five small arms can be observed, each with a button-shaped thickening at the end.

When the proximal centriole is found near the round nucleus, the latter appears slightly flattened in front of the centriole. In this area three to five layers of annulate lamellae are frequently attached to the nuclear membrane (Fig. 10). In their very regular fenestrations and densities they are mostly accurately aligned with the pattern of the underlying nuclear pores (Figs. 10a, 11b). In favorable sections one can also recognize inside the single layers a structure which resembles very closely the diaphragm of the pores in the nuclear envelope (Figs. 10b, c). The ends of the lamellae again frequently pass over directly into the cisterns of the smooth endoplasmic reticulum. At this phase, for the first time a special structure appears in the cytoplasm, a tubular body, the tubules of which have a mean diameter of 16 nm (Fig. 10a). That is about half the distance between the single parallel membranes of the annulate lamellae. The tubules apparently come from a central point and distribute themselves

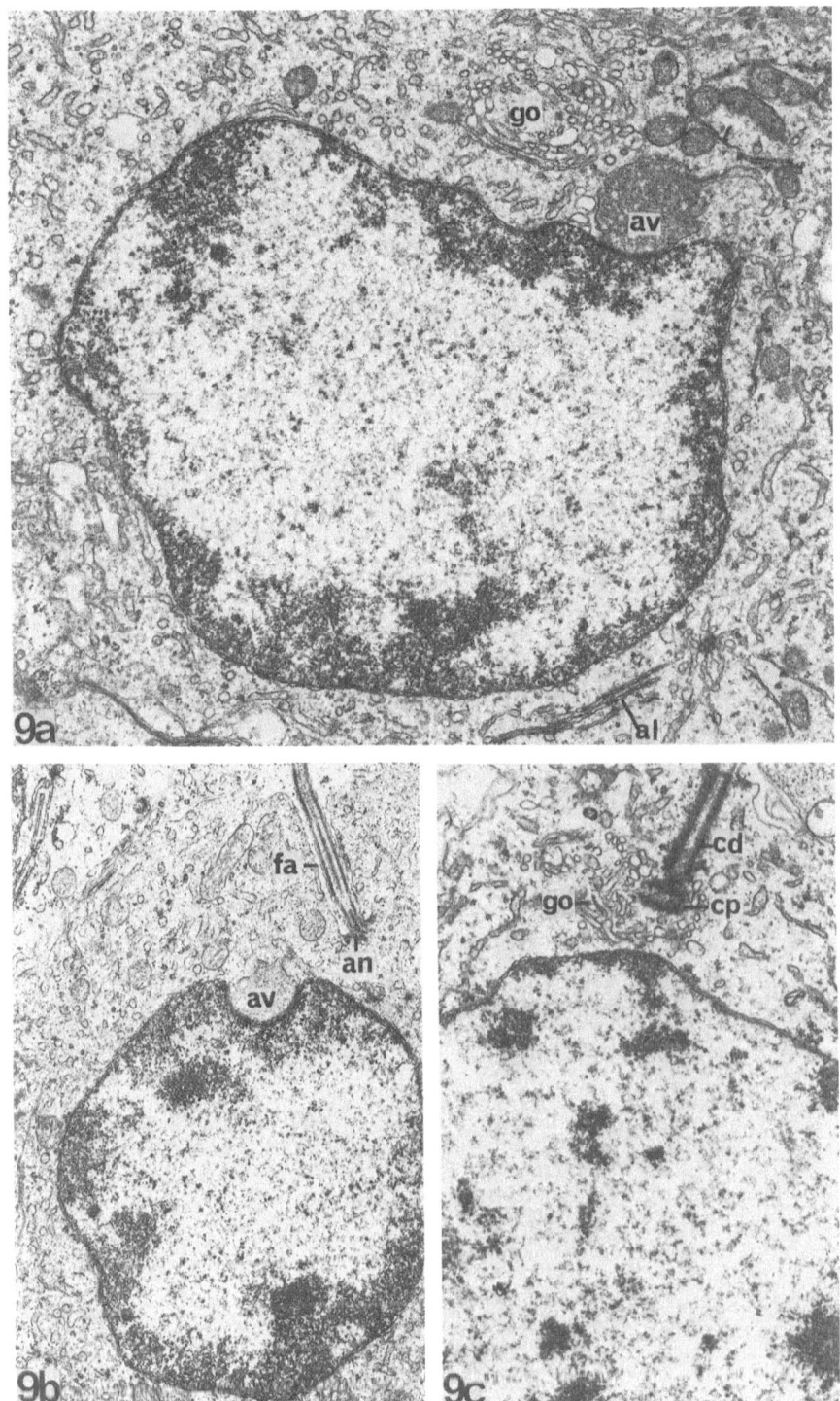

Fig. 9a–c. Early spermatids from the beginning of May. Acrosomal vacuole and centriole have not yet separated. *al*, annulate lamellae; *an*, annulus; *av*, acrosomal vacuole; *cd*, distal and *cp*, proximal centriole; *fa*, flagellum anlage; *go*, Golgi apparatus. *a* × 11 600, *b* × 5800, *c* × 17 000

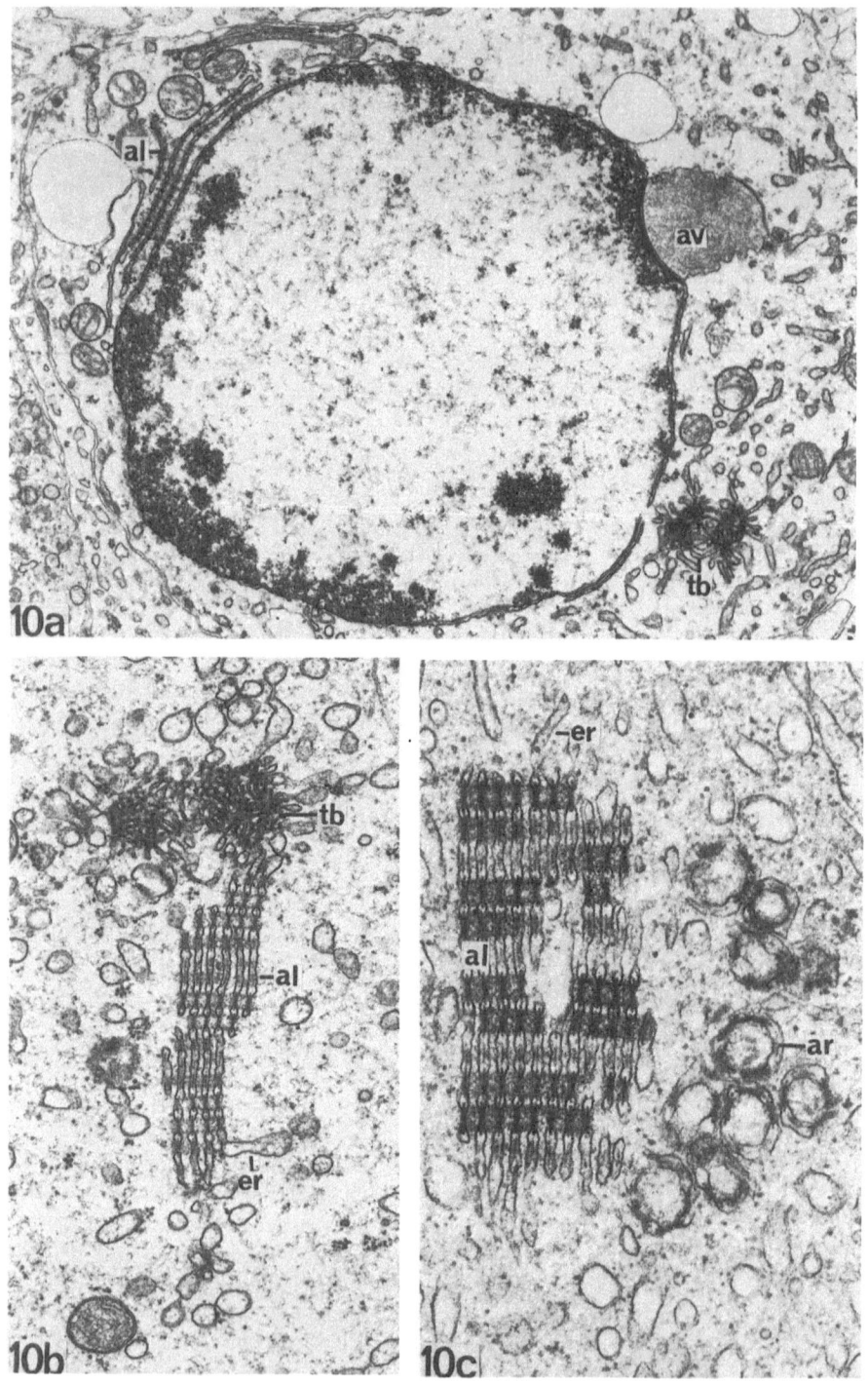

Fig. 10a–c. Spermatids from the beginning of April and May. *a* Early spermatid with the annulate lamellae still in contact with the nuclear envelope. *b* Annulate lamellae connected to a tubular body, from a later spermatid. *c* Annulate lamellae of a later spermatid, some rolled up in form of rings. *al*, annulate lamellae; *ar*, rings of annulate lamellae; *av*, acrosomal vacuole; *er*, endoplasmic reticulum; *tb*, tubular body. *a* × 15 900, *b* × 25 600, *c* × 60 000

in all directions. They seem to be fenestrated similarly to the annulate lamellae, but more irregularly.

The one or two chromatoid bodies present in the cell have no more contact with the nuclear membrane and are mostly to be found some distance from the insertion spot of the proximal centriole. Besides this, an increasing number of multivesicular bodies appear in the cytoplasm.

When the distal centriole has reached its final length of seldom more than 1.2 μm, the proximal centriole has not yet touched the nucleus. Because the whole complex is shifted in the direction of the nucleus, the cell membrane in the area of the annulus is necessarily also pulled towards it. In this way a canal outside the cell develops, which contains the flagellum (Figs. 9c, 11b).

Where the proximal centriole finally touches the nucleus, the nuclear membrane exhibits a hemispherical indentation and is strengthened from inside the nucleus by electron-dense material with simultaneous closing of the nuclear pores, as it occurs similarly in the area of the acrosome vacuole (Fig. 11b). On the side of the cytoplasm the indentation is covered with a thin layer of homogeneous material in which the arms of the proximal centriole are embedded. This cap may have arisen through the melting of the no longer recognizable button-shaped ends of the arms (Fig. 14a). Near the proximal and the primary parts of the distal centriole repeatedly groups of vesicles (Fig. 11b) are to be observed, which probably originate from the nearby Golgi apparatus. Moreover, there are aggregations of contrasting granular material, which one can only suppose come from the chromatoid bodies.

With the insertion of the proximal centriole into the nucleus, a new section of spermatid differentiation begins. At first the acrosomal vacuole, fixed to the nuclear envelope, and the attaching area of the axoneme separate and find themselves on opposite sides of the nucleus. With this the anterior pole of the spermatids is determined, and this is directed with the acrosomal vacuole towards the periphery of the tubule. When the acrosomal vacuole has reached its final position, the nucleus undergoes a noticeable lengthening. As the cytoplasm itself does not expand at first, the nucleus must temporarily take on a coiled shape (Figs. 11a, 12). The side of the acrosomal vacuole opposite to the nucleus thus comes partly into direct contact with the cell membrane (Fig. 11a). Its flocculent contents have only insignificantly increased in density. Simultaneously in the center of the nuclear indentation, under the acrosomal vacuole, a second pipe-shaped indentation of the nuclear membrane occurs. Inside this indentation the next thing to be seen is a rod-shaped mass of electron-dense material which, however, is not surrounded by its own membrane (Figs. 11a, c). In the literature it has often been described as a perforatorium. Its origin is uncertain, because the acrosomal vacuole as well as the nucleus are shut off against this indentation. However, the nuclear membrane in the area of the canal is not thickened by electron-dense material, unlike the area of contact with the acrosomal vacuole (Fig. 13c), and what is more, even nuclear pores can appear. The fine granular chromatin also remains in this phase mostly around the periphery of the nucleus and in some single spiral-shaped threads (Figs. 11a, c), the contrast of which is sometimes increased.

The matrix of the cytoplasm has in the meantime become denser, and the vesicles of the smooth endoplasmic reticulum and the round mitochondria have positioned themselves some distance caudal from the acrosome. Near the insertion of the flagellum, now as earlier, layers of annulate lamellae (Fig. 11b) and also a clearly definable Golgi apparatus can be seen. For the first time now one can recognize in the cytoplasm

31

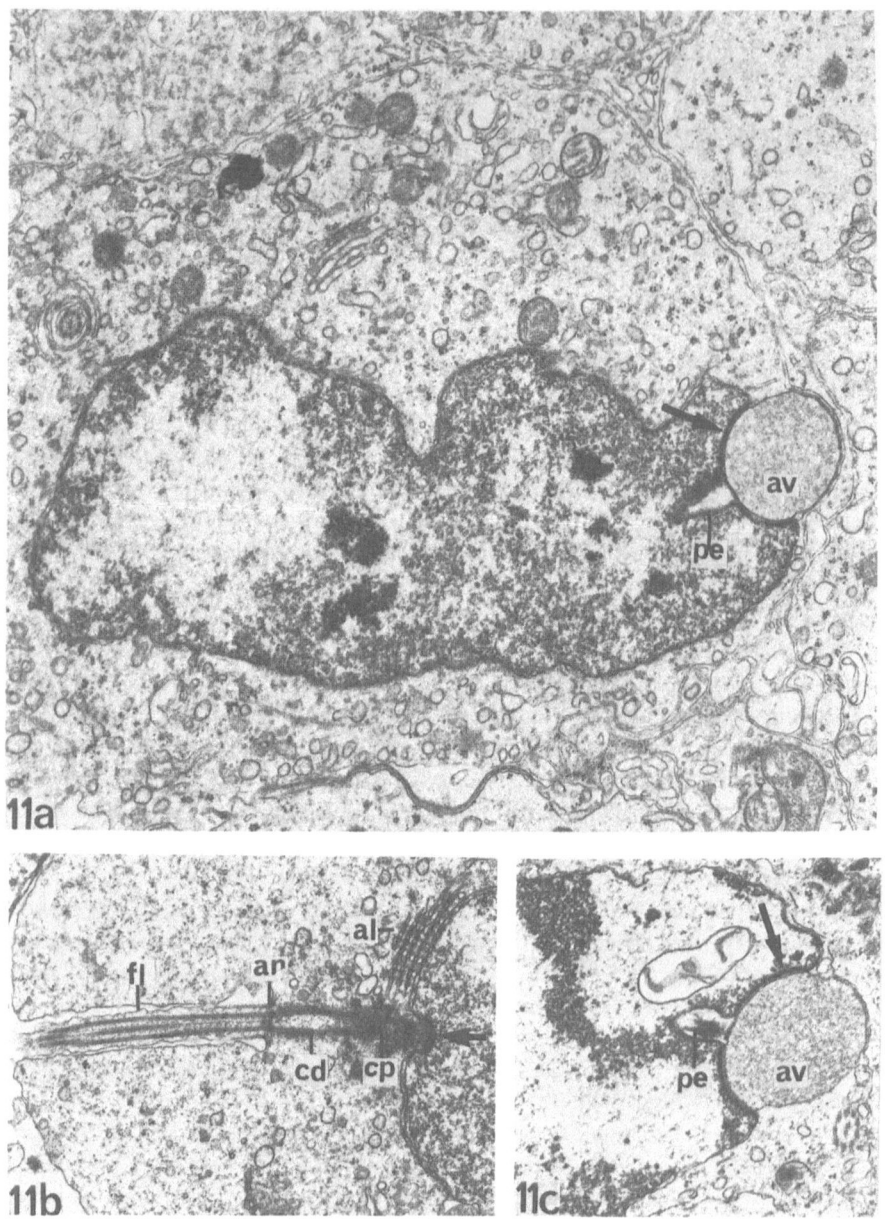

Fig. 11a–c. Spermatids at the start of nuclear elongation from the middle of April and the beginning of May. *al*, annulate lamellae; *an*, annulus; *av*, acrosomal vacuole; *cd*, distal and *cp*, proximal centriole; *fl*, flagellum; *pe*, perforatorium; *arrows*, strengthening of nuclear membrane. *a* × 14 400, *b* × 15 200, *c* × 17 400

microtubules arranged irregularly along the nuclear membrane or a little distance from it. They run transversely to the longitudinal axis of the nucleus and have a diameter of approximately 17 nm. They are missing at both apices of the nucleus in the area of the acrosome as well as around the insertion of the flagellum. Here and there the

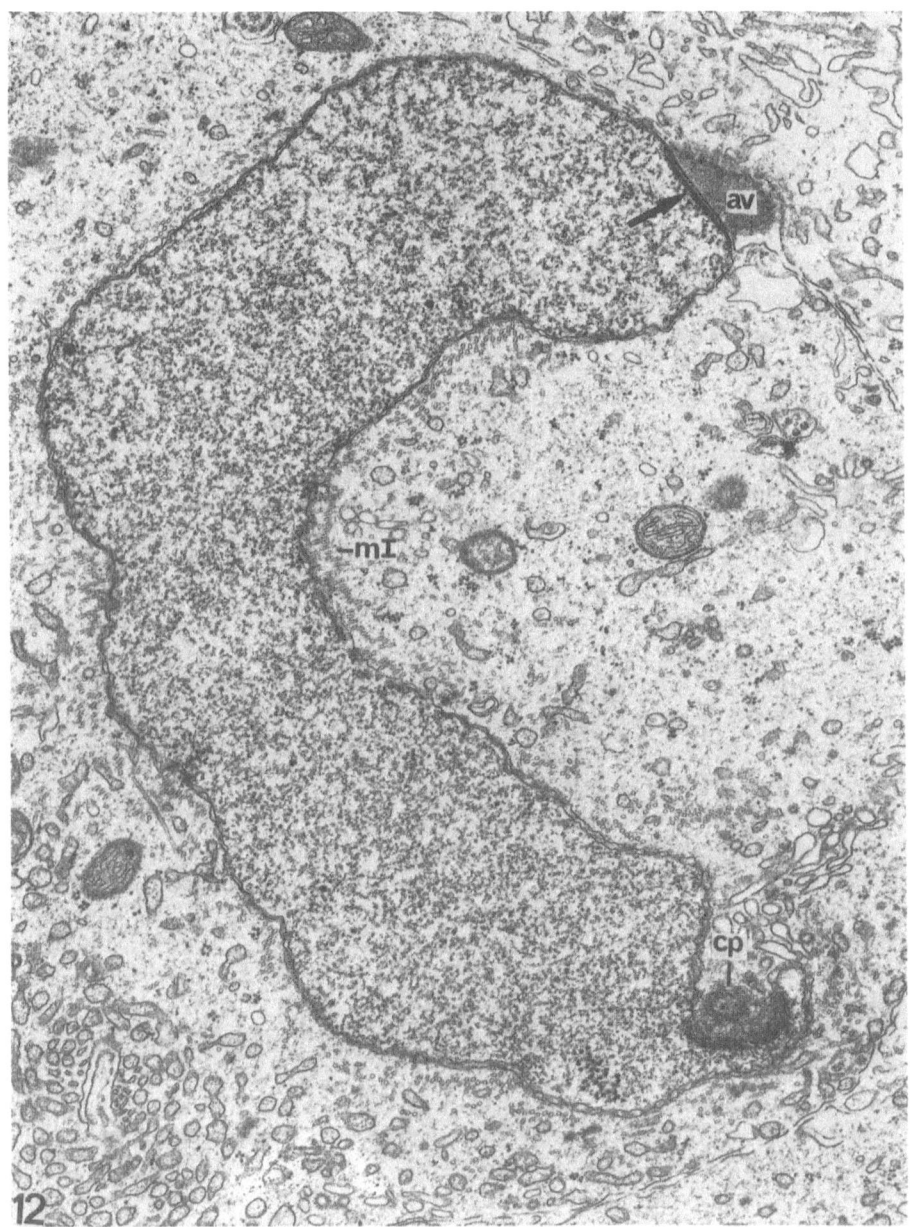

Fig. 12. Spermatid starting nuclear condensation at the beginning of April. *av*, acrosomal vacuole; *cp*, proximal centriole; *mI*, first generation of microtubules; *arrow*, strengthening of nuclear membrane. × 22 700

filamentous-granular complex already seen in the spermatogonia is still to be observed at this time, but without external rough granular components. In the same way single cuts of cilia, independent of the spermatid flagellum, sometimes occur in the cytoplasm, as already described in the secondary spermatocytes.

The next step in differentiation to be observed is a change in the karyoplasm. The fine-filamentous chromatin appears again evenly distributed over the still coiled nucleus, the contrast of which increases gradually (Fig. 12). The elimination of liquid from the nucleus over large vacuoles is, however, not determinable. The increase in contrast depends on the fact that the chromatin has become fine-grained and that these granules, now with an average diameter of 25 nm, have assembled closer. When the nucleus has reached this state, the contents of the acrosomal vacuole have also become denser and have a very homogeneous appearance. In the cytoplasm, the profiles of the smooth endoplasmic reticulum and the multivesicular bodies have increased in number (Fig. 12). The Golgi apparatus lies, as before, near the connection spot of the flagellum. On the other hand the stack of annulate lamellae is now found mostly some distance from the flagellar structures in the cytoplasm, without contact with the nuclear membrane (Fig. 12). Moreover, one can see not only that the number of layers has increased — they vary between 5 and 12 — but also that the annulate lamellae are frequently found in contact with the tubular body (Fig. 10b). Transitions from lamellae into tubules can be observed in quantity, and especially if the tubular body lies in the middle of the lamellar stack and divides the latter into two sections. Moreover, the outermost layers of the annulate lamellae often appear to be rolled into single rings which detach themselves from the array. These rings are also composed of fenestrated double membranes and are not to be mixed up with the usual cross sections of the annulate lamellae (Fig. 10c).

Also striking is the large increase in number of the 17-nm thick microtubules, which can be seen mostly in cross section in one or two rows along the nuclear membrane (Fig. 12). Transverse sections through the spermatid show that they run around the nucleus in very flat helices as so-called manchette, which only begins some distance from the acrosome and barely reaches the caudal end of the nucleus.

Before the actual condensation of the nucleus begins, one finds the tubules of the manchette very regularly ordered in a single row from the acrosome to approximately the level of the distal centriole (Fig. 13b). In the following steps of development the whole cell together with its nucleus appears to become more and more elongated. Thus the nucleus becomes simultaneously thinner and finally it is no longer twisted. As a visible sign of the start of nuclear condensation there occur at first single and then increasingly numerous granules or short rods, with an average diameter of approximately 45 nm, until in the end nearly the whole of the nucleus is filled with them. Only beneath the nuclear membrane is there a space free of granules to be seen, which varies in width between 50 and 100 nm (Fig. 13a), and this is retained during further condensation almost until the complete maturing of the spermatid. At first a fine filamentous material is still recognizable between the single granules (Figs. 13b, c), but this can no longer be seen when the granules are more closely pressed together.

With the start of the nuclear condensation the development of the acrosome continues simultaneously. The rod which extends in a canal under the acrosomal vacuole some distance into the nucleus now also continues outside the nucleus in a cranial direction. The acrosomal vacuole thus becomes centrally detached from the nucleus (Figs. 13a, c). Peripherally, however, the contact between the nucleus and the acrosomal vacuole remains. Also in this area the nucleus is pushed cranially upwards and encircles the central canal in the form of a cushion. Only this approximately 0.4-μm high swelling is covered by the acrosomal vacuole on the outside, so that the space which contains the rod is shut off with a lid (Fig. 13c). This means that the acrosome

34

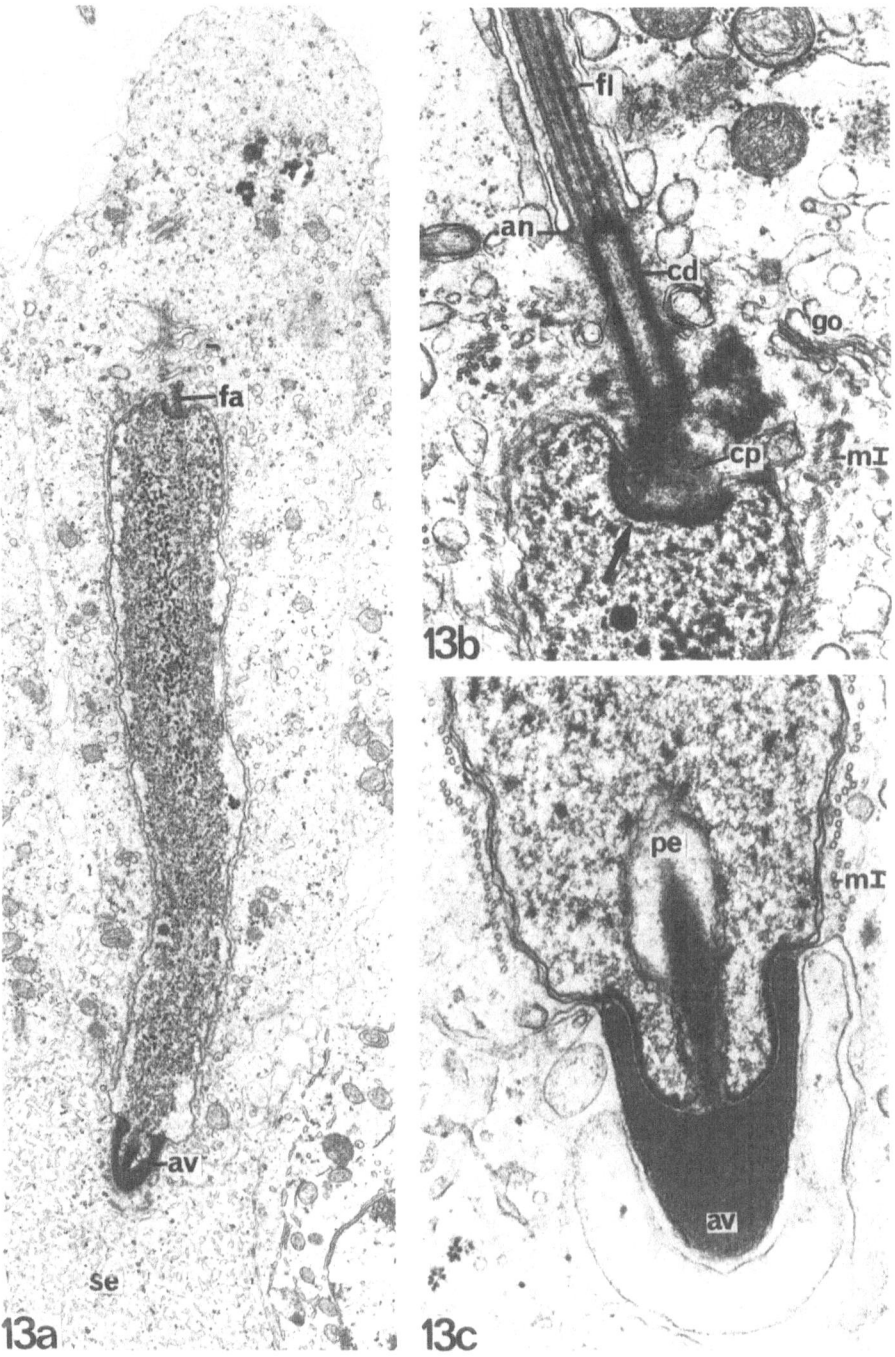

Fig. 13a–c. Spermatids during nuclear condensation at the middle of April and the beginning of May. *a* At the beginning of condensation. *b* The nuclear membrane in the area of insertion of the proximal centriole is strengthened from both sides, nuclear and cytoplasmic. *c* Lid-like closing of the nuclear indentation. *an*, annulus; *av*, acrosomal vacuole; *cd*, distal and *cp*, proximal centriole; *fa*, flagellum anlage; *fl*, flagellum; *go*, Golgi apparatus; *mI*, first generation of microtubules; *pe*, perforatorium; *se*, Sertoli cell; *arrow*, strengthening of nuclear membrane. *a* × 8300, *b* × 30 000, *c* × 35 200

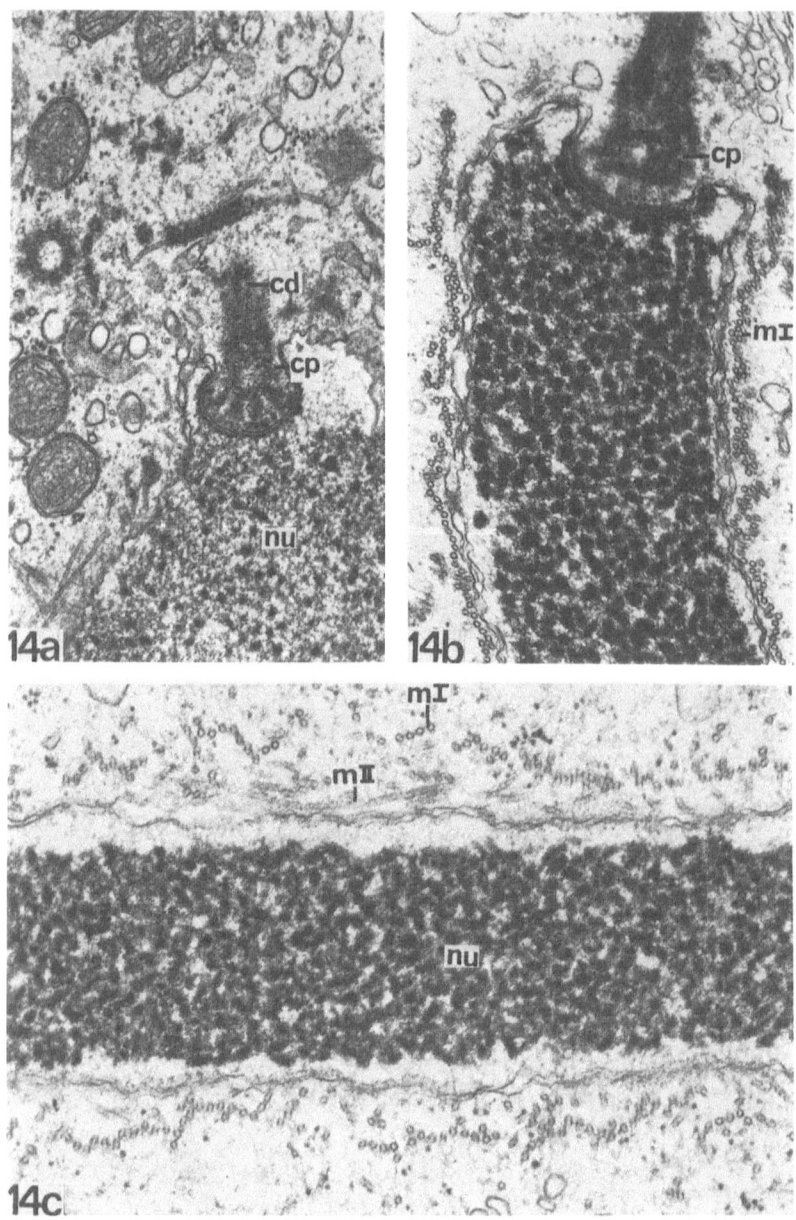

Fig. 14a–c. Sections of spermatids at advanced nuclear condensation from the middle of April and the beginning of May. *cd*, distal and *cp*, proximal centriole; *mI*, first and *mII*, second manchette of microtubules; *nu*, nucleus. *a* × 30 200, *b* × 40 000, *c* × 32 400

does not form a cap over the nucleus, but is situated for the most part cranially from it. The rod inside the canal is frequently split several times (Figs. 13c, 15b).

When the condensation of the nucleus has progressed so far that the approximately 45-nm thick granules fill the whole of the nucleus in a dense package, the nucleus stretches considerably once again. As it becomes thinner in the course of elongation, the chromatin granules touch each other and gradually melt into one another to form

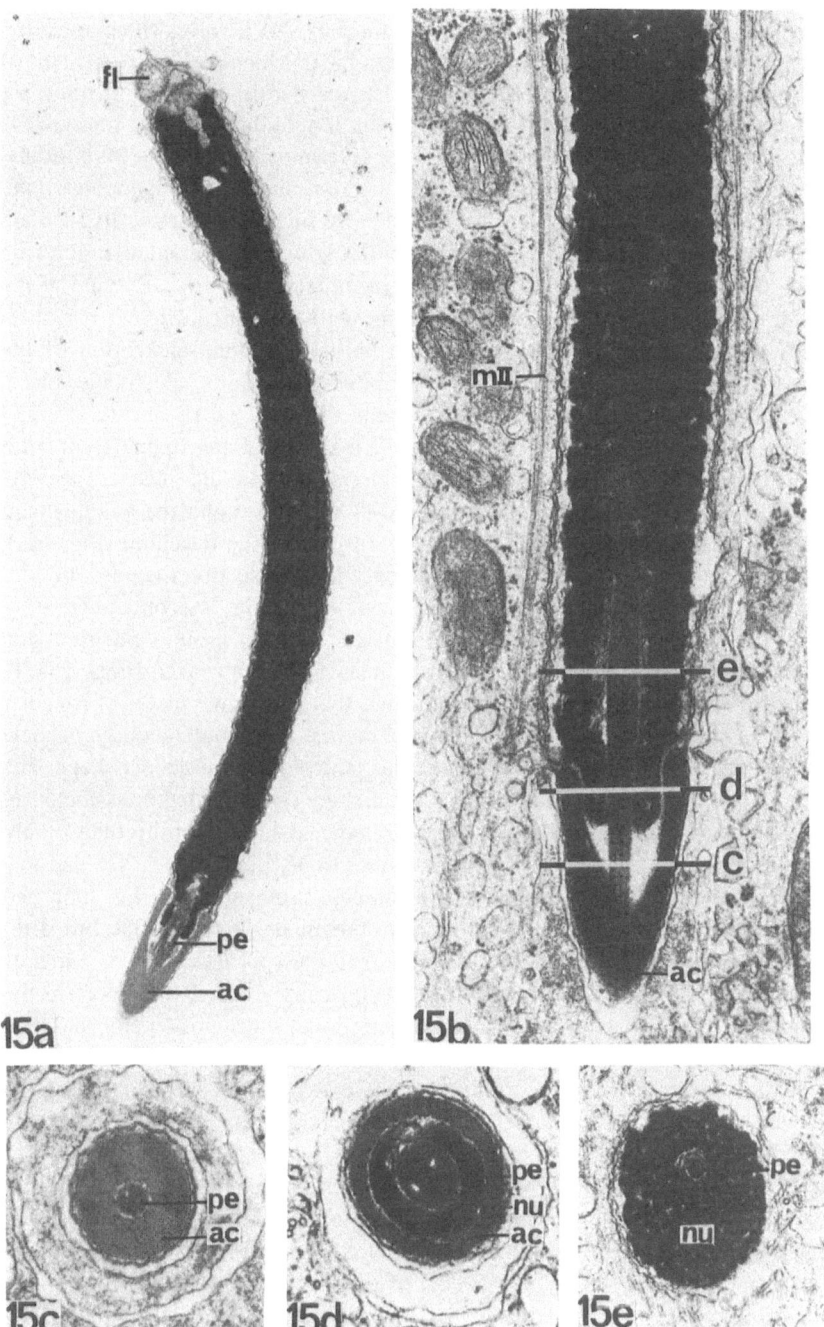

Fig. 15. *a* Section of spermatozoon from the tubular lumen, *b* longitudinal and *c-e* cross sections of the apical region of almost mature spermatids at the beginning of April and May. The planes of section in *c-d* are marked in *b*. *ac*, acrosome; *fl*, beginning of the flagellum; *mII*, manchette of longitudinally oriented microtubules; *nu*, nucleus; *pe*, perforatorium. *a* × 13 100, *b* × 27 800, *c, d* × 37 800, *e* × 25 500

larger units (Figs. 14b, c). The stretching of the nucleus also involves the manchette. The result is a change in orientation of the microtubules which previously ran almost circularly round the nucleus. When the nucleus has reached its final length, they are oriented in a direction more or less parallel to the longitudinal axis of the nucleus (Fig. 15b). Positions intermediate to the strictly determined circular or longitudinal course of the microtubules occur comparatively seldom. Similarly both circularly and longitudinally running tubules can be observed side by side only sporadically, and are confined to rather small areas. In this case the longitudinally oriented tubules lie between the circular manchette and the nuclear membrane (Fig. 14c). Because of the missing intermediate stages, one cannot determine with certainty whether it is a re-arrangement of the microtubules in connection with the sudden elongation of the nucleus, or the development of two successive tubular generations, which takes place. In both directions the microtubules have the same diameter of 17 nm. After final elongation the nucleus has a length of 12 μm and a diameter of approximately 0.6 μm (Fig. 15a). The manchette of the longitudinally running microtubules reaches from the acrosome (Fig. 15b) to the caudal end of the cell, that is beyond the annulus, and accompanies the canal which contains the proximal part of the flagellum (Fig. 16a). Because of the decrease in the nuclear circumference in cross section the microtubules appear very closely to each other and must frequently escape into a second layer.

During advancing differentiation the karyoplasm becomes more condensed, and when the terminal stage is reached no substructures can be recognized (Figs. 15a, b, 16a–c). Nuclear vacuoles are absent; now and then there are some places of irregular and incomplete condensation. The cytoplasmic matrix has simultaneously become much denser (Fig. 4), and on the whole the cell has assumed a more slender shape. The mitochondria and the smooth endoplasmic reticulum are concentrated predominantly in the caudal part of the spermatid. The annulate lamellae have given up their rigidly packed arrangement, but there are still groups of rings to be found.

In the area of the distal centriole the electron-dense substance which is supposed to originate from the chromatoid body is recognizable in single collections, but without contact with the annulus (Figs. 16a, b). This material appears at last in a row of granules attached to the centriole from outside (Fig. 16a). The proximal core in the hollow cylinder of the distal centriole is then reduced to a short closing-plug and likewise a small piston extends distally in the hollow cylinder (Fig. 16c). Next to this the triplet formation of the centriole continues in the characteristic substructure of the axoneme, which begins cranial from the annulus, i.e., still in the area of the definite middle piece (Figs. 16a, b). The axoneme consists of two single central tubules and nine peripheral doublets, of which one tubule is always empty and the other electron-dense (Figs. 16f–i). In this part up to the beginning of the definitive principal piece, the peripheral doublets are centrally and peripherally accompanied by an electron-dense substance which can merge into a sort of ring on the outside. Because of the irregular distribution of this material, sometimes even between the pairs of tubules, the typical dynein arms of the dense tubules are mostly hidden in this part of the tail. Centrally the material looks like single fat granules, in both transverse and longitudinal sections (Figs. 16b–d). Then at the level of the annulus the fibrous sheath begins, which surrounds the axoneme, separated by a light cleft, along the whole of the principal piece, first as a complete, more distally as an intermittent ring (Figs. 16f–h). It is formed relatively late, as the condensation of the nucleus and the cytoplasm is already far advanced. The structure of the tail alters in the distal direction, so that first the

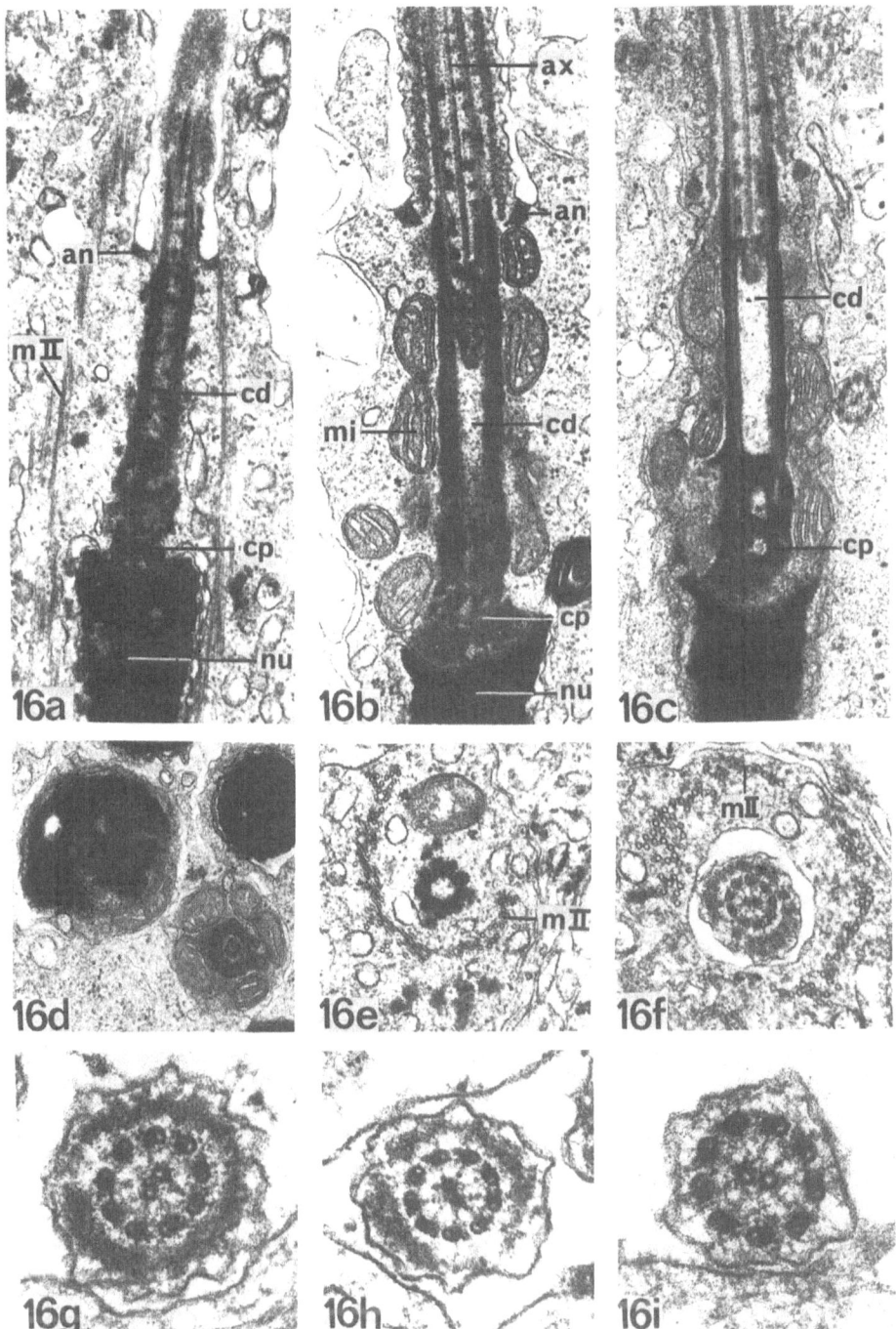

Fig. 16a–i. Transition from the nucleus to the flagellum in mature spermatids at the middle of April and beginning of May. *a–c* Longitudinal sections of the distal end of the nucleus, middle piece and beginning of principal piece. *d–i* Sequence of cross sections of the flagellum in distal direction through the proximal centriole (*d left*), through the lower (*d* bottom *right*) and the middle (*e*) region of the distal centriole and through the principal piece (*f–i*). *an*, annulus; *ax*, axoneme; *cd*, distal and *cp*, proximal centriole; *mII*, second microtubular generation; *mi*, mitochondrium; *nu*, nucleus. *a, b* × 30 000, *c* × 29 300, *d* × 26 700, *e* × 25 100, *f* × 32 600, *g* × 68 400, *h* × 65 600, *i* × 82 800

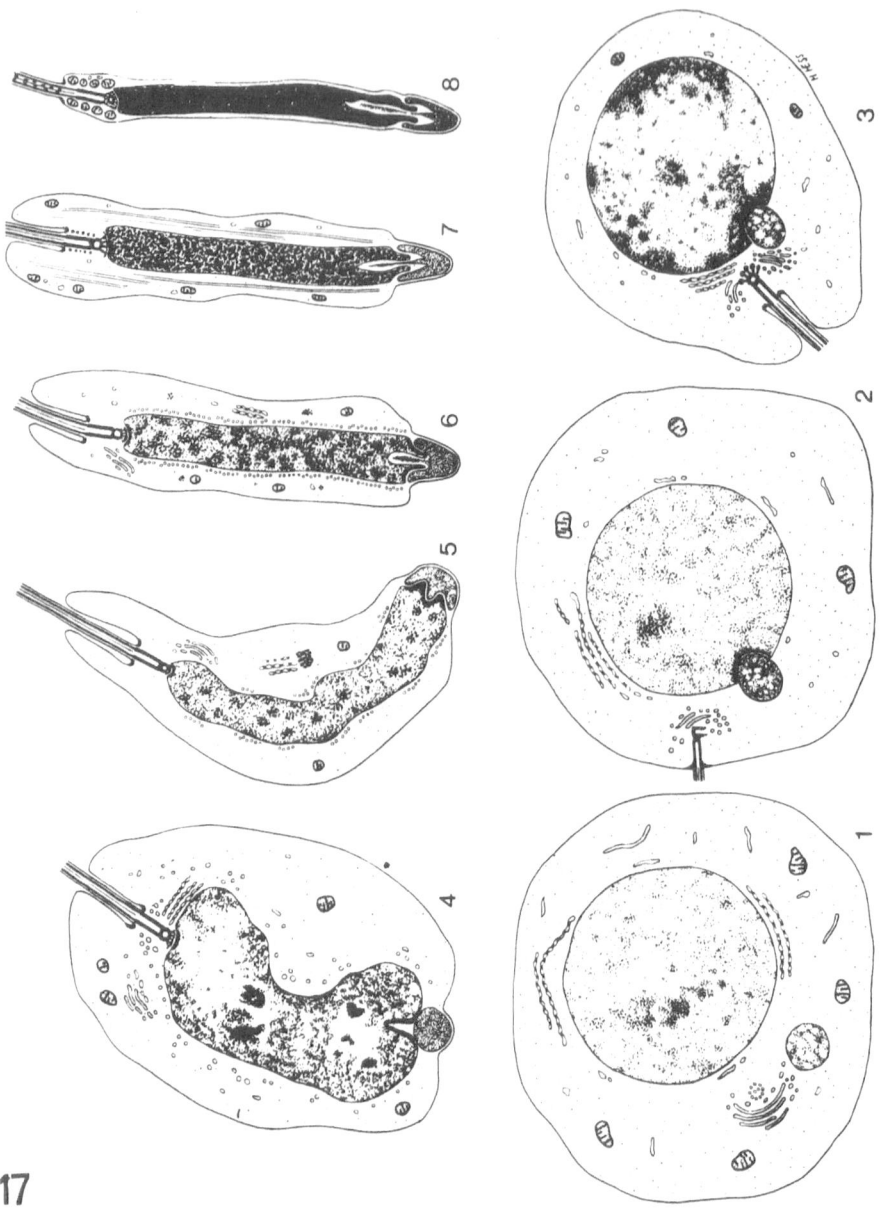

Fig. 17. Diagrammatic representation of spermatid differentiation in the swan. Numbers indicate the chronological sequence of steps of differentiation and the direction towards the basal lamina

fibrous sheath terminates, then the compact tubules of the axoneme become optically empty, and finally in the end piece the regular 9 + 2 order is given up.

At the last differentiation step of the development of the spermatid the final formation of the middle piece occurs. The mitochondria are continuous to the nucleus ordered round the proximal and distal centrioles (Figs. 16b–d), so that a marked neck piece is missing. In transverse sections there are four or five mitochondria as a rule

(Fig. 16d); the same number are also found in the longitudinal sections, with an average of five mitochondria on each side of the centrioles. With the release from the germinal epithelium, the spermatids abandon their cytoplasm, so that when moving freely in the lumen of the seminiferous tubule they carry no more drops of cytoplasm with them (Fig. 15a). However, one can recognize the annulus, now as before, at the end of the middle piece of these spermatozoa. The whole process of spermatid differentiation is summarized diagrammatically in Fig. 17.

6.5 Sertoli Cells

The Sertoli cells, the only somatic cells which are present in the seminiferous tubules along with the different types of germ cells, can show very differing pictures depending on their functional state. During the active phase of spermatogenesis, they are not at their peak of activity, so that at this time they appear relatively inconspicuous. Nevertheless one can differentiate them from the germ cells through certain characteristic structures. The Sertoli cells always have contact with the basal lamina (Fig. 6), and on the other hand they extend upward to the center of the tubule. Their cell bodies are shaped like stars and branch out quite far; with these branches they stretch between the germ cells. Frequently they underlie the spermatogonia over a larger area along the basal lamina, as a very thin layer of cytoplasm (Figs. 5, 6). This can only be seen by means of electron microscopy, so that in light microscopic pictures the false impression occurs that the spermatogonia lie with a broad base on the basal lamina. The Sertoli cells are in direct connection with each other through their processes, but without special cell contacts being developed. One rather finds junctions between Sertoli cells and spermatogonia or primary spermatocytes, where collections of electron-dense material are symmetrically adjoined to the cell membranes (Fig. 7a). Laterally the Sertoli cells exhibit interdigitations which grow in number from the base towards the lumen and through which the cells are linked together. But also they repeatedly surround the germ cells and especially the spermatids during advanced differentiation. In particular the heads of nearly mature spermatids are embedded between the Sertoli cells, but without special structures such as filaments or tubules in the Sertoli cells for anchoring (Fig. 13c).

Especially typical of the Sertoli cells is their nucleus, which is seldom found in the base layer but mostly in the second or third layer of the germ cells, that means in between the primary spermatocytes (Figs. 4, 6, 18b). It is round and has no indentations (Figs. 4, 6, 18a, b) and has a diameter of 7.5—8 μm. At the inner layer of the nuclear membrane lies a row of contrasting granules which makes it appear thicker than the outer layer. The chromatin is fairly evenly distributed over the whole nucleus and is embedded in a matrix, which is more electron-dense than that of the germ cell nuclei. The nucleolus, typically for the Sertoli cells, is composed of three components: granular, fibrous, and amorphous. Near the nucleolus one frequently finds one or two EDTA-positive complexes with the same size of grain as are found in the various types of germ cell nuclei (Figs. 6, 18a).

The most striking constituent of the cytoplasm is the smooth endoplasmic reticulum, which in the form of cisterns and numerous tubules fills the whole cell (Fig. 18a). Sometimes ribosomes, either single or positioned in a row, can be attached to cisterns arranged in parallel (Fig. 6). The ribosomes also occur free in the cytoplasm, mostly as

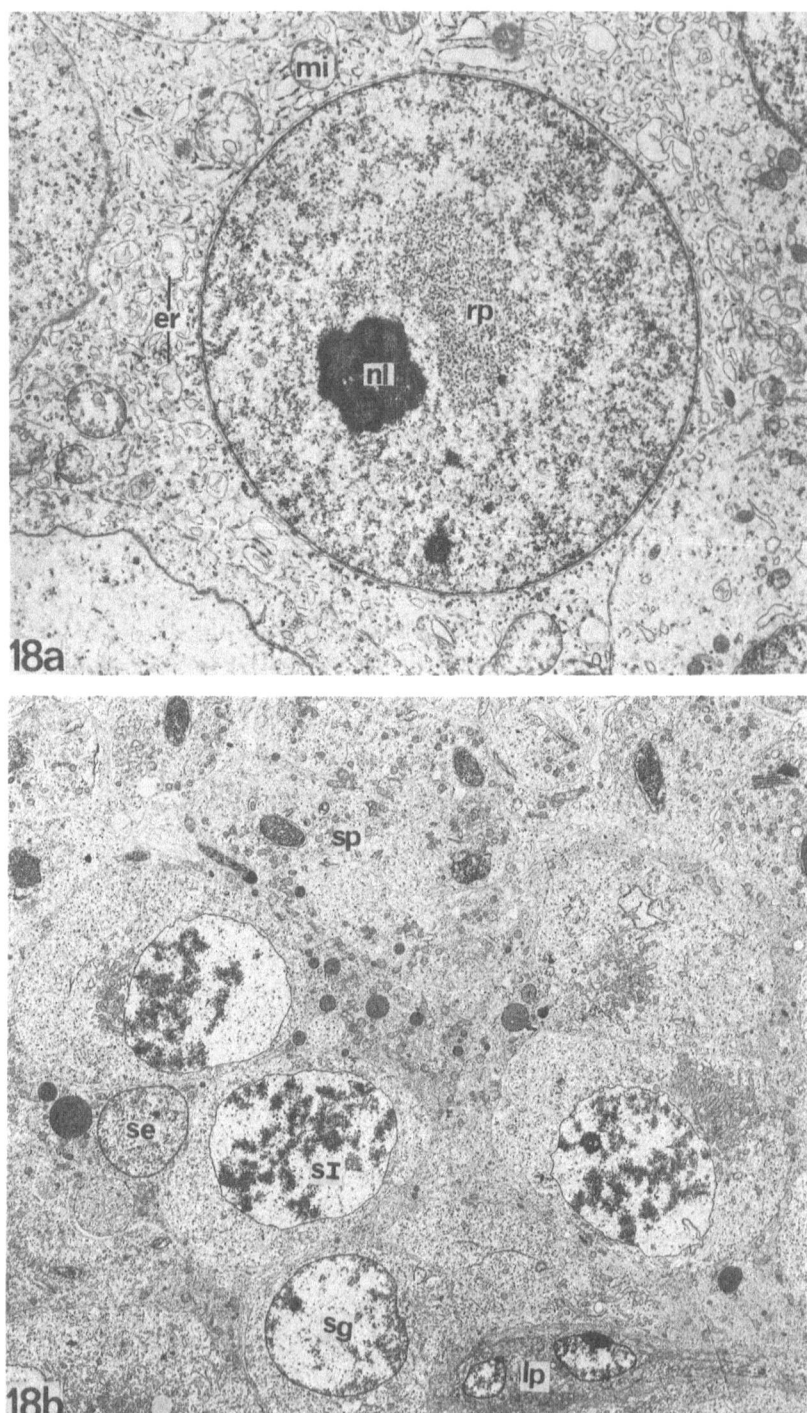

Fig. 18a, b. Part of a Sertoli cell from the middle of April *a* and section of a seminiferous tubule from the beginning of April *b*. *er*, endoplasmic reticulum; *lp*, lamina propria; *mi*, mitochondrium; *nl*, nucleolus; *rp*, ribonucleoprotein complex; *se*, Sertoli cell; *sg*, spermatogonium; *sI*, primary spermatocyte; *sp*, spermatid. *a* × 9000, *b* × 2350

polysomes. A Golgi apparatus is not always clearly to be seen. The mitochondria are round or oblong; they contain clearly recognizable cristae in their dense matrix and are usually somewhat bigger than those in the neighboring germ cells. Lipid drops are only very seldom present.

Landmark features of the Sertoli cells are the differing densities of the cytoplasmic matrix and the changing amount of inclusion bodies (Figs. 4, 6, 18b), which possibly express different functional stages. In the lighter cells which are more often present, there are additional single microtubules between the profiles of the endoplasmic reticulum and thin filaments in longitudinal or transverse sections, occasionally joined together in bundles. They are presumedly hidden when the cytoplasmic matrix becomes denser. Then simultaneously the pleomorphous osmiophil inclusions with very differing contents increase in number. In general they are surrounded by a membrane, so that it could be supposed that they were lysosomes or the remains of residual bodies. They are present near the nucleus, mostly supranuclear. In the vicinity of the lysosomal inclusions one sometimes finds remains of degenerated cells enclosed in a membrane. Only in the case of primary spermatocytes with synaptonemal complexes, however, can one see clearly to which type of germ cell they belonged.

The characterization of the Sertoli cells given here is principally valid for the timespan during which the germinal epithelium is fully active. During the regressive and the resting periods of the seminiferous tubules the cells can show a somewhat different picture, which will be described later. The different states of activity of the Sertoli cells during the active period of spermatogenesis, however, are not related to any state of development of the germ cells, such as the different stages of spermatogenesis, for example. Therefore even Sertoli cells with different densities and numbers of inclusions can be in direct contact with one another over a large area.

6.6 Lamina Propria

The lamina propria of the seminiferous tubule is made up of the basal lamina plus a lamination of peritubular cells and extracellular fibrous material (Fig. 19a). The basal lamina follows in its course the basal cell membrane of the spermatogonia and Sertoli cells, and can occasionally be split over short stretches. Next to it on the outside is a layer of mostly collagenous fibrils with no favored alignment. The first layer of cells after that is made up of fibroblasts which are arranged in a circle round the tubule. The next one to several single layers of cells also run in a circle round the tubule with always more or less the same intercellular space between them. The second and following cell layers are different from the first in that they contain in their cytoplasm filaments and zones of densities along their cell membranes (Fig. 19a). For this reason they are referred to as "myofibroblasts". In the intercellular spaces one finds fibrous material which can be identified as collagenous fibrils. Additionally one can quite frequently observe unmyelinated nerve fibers between the myofibroblasts. Mostly there are several of them bundled together and surrounded by processes of a Schwann cell (Figs. 19a, b). Inside the axons one finds, in addition to neurotubules and single mitochondria, electron-dense vesicles with a light halo around a dark center. In the larger areas of the interstitium in the angular interstices between the tubules these nerve fibers are in contact with blood-vessels or Leydig cells.

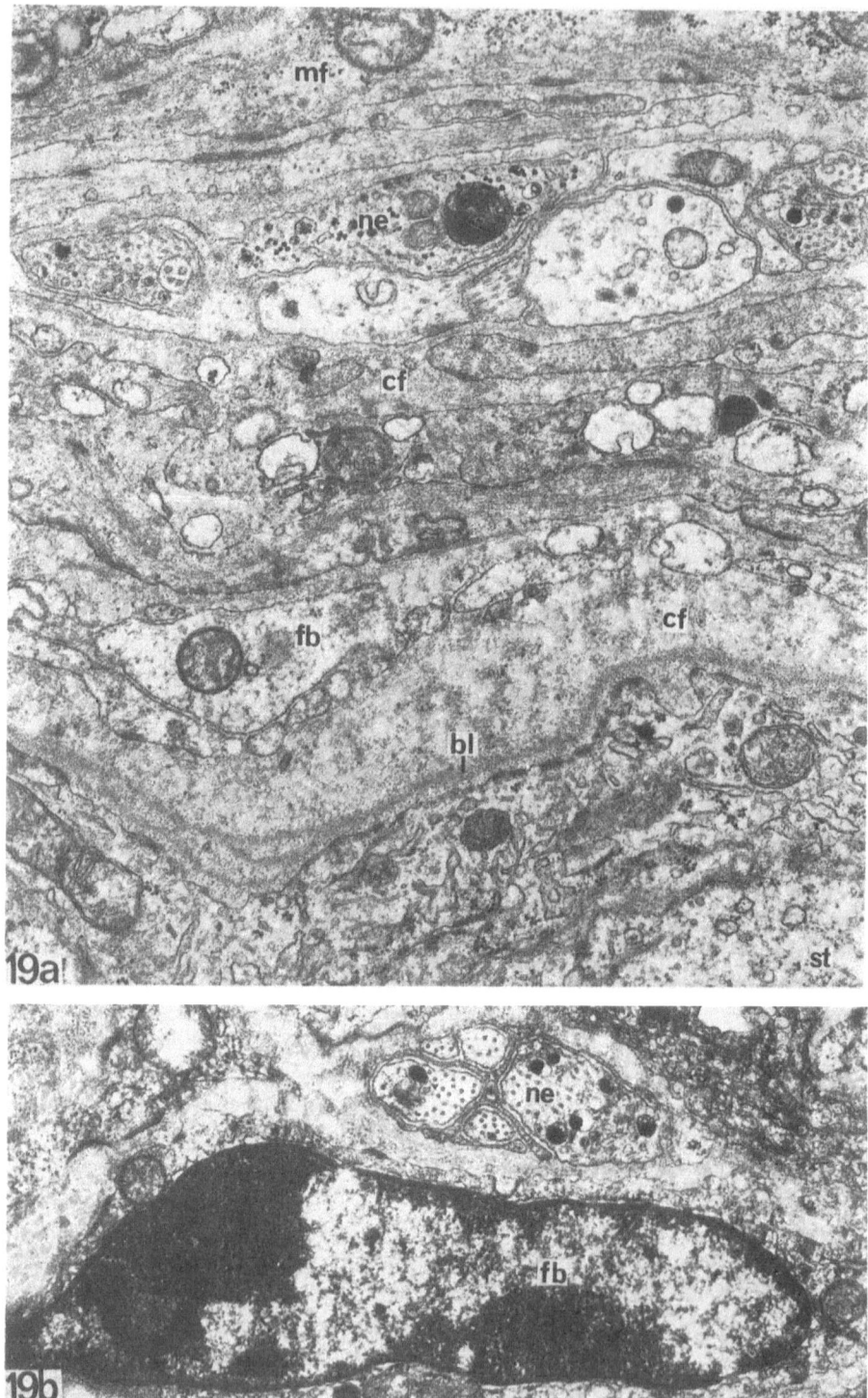

Fig. 19a, b. Section from the lamina propria at the beginning of April *a* and the end of August *b*.
bl, basal lamina; *cf*, collagenous fibers; *fb*, fibrocyte; *mf*, myofibroblast; *ne*, nerve fiber; *st*, seminiferous tubule. *a* × 25 700, *b* × 22 700

6.7 Discussion

Electron microscopic investigations of spermatogenesis in birds have so far dealt rather exclusively with the development and differentiation of single cell structures. A documentation of the ultrastructure of the complete process of male germ cell development, from spermatogonium to mature free-moving spermatozoon, is available only in the work of Marchand (1977), who described the process of spermatogenesis in the domestic Muscovy duck (*Cairina moschata* L.). Like the swan, this species belongs to the Anatidae family (Grzimek et al. 1968), although to a different subfamily (see p. 5), and it therefore offers especially good possibilities for comparison.

The germ cells lying next to the basal lamina, the spermatogonia, are characterized both in the swan and in the duck by their round or kidney-shaped nuclei, which are very poor in chromatin. The granular ribonucleoprotein complex of about the same size as the nucleolus, which is present in most of the nuclei in the swan, can frequently be observed also in the drake, but no satisfactory statement on its meaning can be made.

Particularly noticeable features in the cytoplasm are the mitochondria, mostly grouped on one side of the nucleus, and between which there is no intermitochondrial cementing substance, and the filamentous-granular body. This is presumably identical with a complex which in the duck is termed "chromatoid body" even in the spermatogonia (Marchand 1977). In its substructure it is similar to the nucleolus-like body or the nuage of primordial germ cells, which have been described by Kalt (1973) in *Xenopus* and by Eddy (1974) in the rat. In the swan the spermatogonia represent a polymorphous cell generation, of which the single components are present in varied distribution and independently of one another. Therefore a differentiation of different cell types as it is known in mammals (Rowley et al. 1971; Clermont 1972) is not possible. Marchand (1977) points out the same difficulties in the duck. Even a differentiation between stem spermatogonia and those cells which develop further to give rise to spermatocytes is impossible. Features such as a broad-based contact with the basal lamina or intercellular bridges, which in *Xenopus* characterize B-spermatogonia (Kalt 1973), are relatively seldom found in the swan and are therefore insufficient as morphological criteria. The EDTA-positive body, on the other hand, is a structure which can be observed in the nuclei of nearly all spermatogonia and moreover it is not confined to these, so it cannot be compared with those nuclear inclusion bodies which were described as characteristic for A-spermatogonia in human material (Tres and Solari 1968).

During the single stages of the first meiotic prophase the development of the primary spermatocytes in the swan is similar to that in the duck (Marchand 1977). Particularly noticeable is the more or less heavy compression of the chromatin during zygotene, and its positioning on that side of the nucleus where simultaneously in the cytoplasm the mitochondria, the Golgi apparatus, and the centrioles concentrate. However, there is no sign to indicate that there is an enhanced exchange of substances between nucleus and cytoplasm during this phase. A similar aggregation of mitochondria on one side of the nucleus without the simultaneous chromatin positioning is demonstrated photographically by Nagano (1959) in the spermatocytes of the cock. In the swan too the pairing of the homologous chromosomes in zygotene and pachytene finds its morphological expression in the formation of synaptonemal complexes, as first described by Moses (1956, 1958) in a crab and a salamander. Their tripartite fine

45

structure and their close contact with the nuclear membrane are fundamental characteristics of the chromosomes of meiotic prophase in both male and female germ cells. Synaptonemal complexes have been found in numerous species of animals and plants, including the pigeon (Fawcett 1956; Nebel and Coulon 1962), the cock (Coleman and Moses 1964), and the duck (Marchand 1977).

A further characteristic structure found in the primary spermatocytes of the swan are the one or more chromatoid bodies. These are to be seen for the first time in pachytene, predominantly in contact with the nuclear membrane, which in this area exhibits an increase in nuclear pores. In this the chromatoid body corresponds to findings in various mammals (Fawcett et al. 1970). Also in its fibro-granular substructure it is similarly heterogeneous, but in the swan the vesicles which can be observed in mammals (Russell and Frank 1978) and the additional satellite are missing (Fawcett et al. 1970; Susi and Clermont 1970). Comparable structures have also been found in spermatogonia and spermatocytes of amphibians (Kalt 1973), fish (Schjeide et al. 1972), and even invertebrates (Reger et al. 1977).

There are several different theories about the origin of the chromatoid body. In mammals it is thought either that the chromatoid body is formed from material from the cell nucleus (Comings and Okada 1972; Söderström and Parvinen 1976), or that the intermitochondrial cementing substance in the spermatogonia as material from the cytoplasm represents a precursor of the chromatoid body (Fawcett et al. 1970; Beams and Kessel 1974). In some cases the chromatoid body is seen as a special form of nuage in the male germ cell, e.g., in the rat (Eddy 1974; Russell and Frank 1978), and would thus already have its origin in the primordial germ cells, similar to the nucleolus-like body in *Xenopus* (Kalt 1973) which in this species becomes a definite chromatoid body. Since in the swan there is no intermitochondrial cementing substance in the spermatogonia, this cannot be the preliminary material of the chromatoid body. On the other hand, in the spermatocytes going through pachytene a close relationship between the chromatoid body and the fibrous-granular complex already present in the spermatogonia can often be found, so it is possible to surmise that this complex has something to do with the formation of the definitive chromatoid body.

The secondary spermatocytes are characterized in many animal species, particularly mammals, by the garland-shaped circular arrangement of the smooth endoplasmic reticulum in contact with mitochondria. These garlands form in the swan already by the end of prophase in the primary spermatocytes and are thereby especially obvious in the wide intercellular bridges during the division phase. Special membrane systems as observed by Nagano (1961) in the narrowing intercellular bridges during both telophases of the cock cannot be found in the swan.

With great regularity one finds in the secondary spermatocytes of the swan those membrane arrays which Swift (1956) called annulate lamellae and which are also characteristic of these cells in the duck (Marchand 1977). These structures are a very uniform and widespread component of fast-growing and differentiating cells. They have therefore particularly been observed in the developmental stages of male and female germ cells in numerous invertebrates and vertebrates (see Kessel 1968). Special features are the frequently found direct transitions of the lamellae into the cisterns of the endoplasmic reticulum, as well as their similarity to the nuclear membrane, characteristics which are also true for the swan. Because of the frequent conformity of the annulate lamellae to the nuclear pore system, the lamellae are thought to be derivatives of the nuclear membrane. About their function, e.g., a transmission of information from the nucleus to the cytoplasm (Kessel 1968), only speculations can be made. In the swan the relationship of the annulate lamellae to at least one of the chromatoid bodies in the secondary spermatocytes is remarkable. A similar close contact between these two cell organelles was described in the spermatogonia of a bony fish (Schjeide et al. 1972). In this case the material for the chromatoid body is said to come from the nucleus and to stimulate the formation of the annulate lamellae. No clear statement was made about the function of the lamellae.

The differentiation of spermatids, or at least separate aspects of it, have been dealt with in a restricted number of electron microscopic investigations of various avian species. These observations show that there is a series of conformities in the development in the swan with that in the drake (Marchand 1977; Maretta 1977) and the cock (Nagano 1959, 1962; Okamura and Nishiyama 1976; Gunawardana and Scott 1977; Maretta 1977). These conformities concern the development of the acrosome, the condensation of the nucleus, and the formation of the tail structures.

The acrosomal vacuole, arising in the area of the Golgi apparatus, contains – contrary to the case in countless animal species including man (Holstein and Wartenberg 1976) – no special granule in an optically empty vacuole, but is filled from the beginning with a flocky material, which in equal distribution increases in density. The development starts similarly in the budgerigar (Humphreys 1975a). Whereas in this species the acrosomal vacuole only takes up a ring-shaped contact with the nuclear membrane, in the swan at first a third of its surface touches the nucleus. This is also true in the cock (Nagano 1962) and the drake (Marchand 1977), as well as in different species of pigeons (Mattei et al. 1972; Yasuzumi and Yamaguchi 1977). The primary free subacrosomal space, which in the budgerigar arises directly with the beginning of contact between the acrosomal vacuole and the nucleus (Humphreys 1975a), occurs later in the swan, the drake (Marchand 1977), and the cock (Nagano 1962) in the form of a tube-like nuclear invagination. These initially varying contact zones have their consequences for the later attachment of the acrosome to the nucleus. This connection is extremely unstable in the budgerigar because the acrosome remains cranial to the nucleus, whereas in the swan, drake, and cock a somewhat more stable anchoring takes place through a lid-like overlapping of both structures, without, however the formation of an acrosomal cap as found in the pigeon (Mattei et al. 1972; Yasuzumi and Yamaguchi 1977).

A characteristic structure belonging to the acrosome in ducks and geese is the rod which reaches into the subacrosomal invagination of the nucleus. It is typical also of the cock (Nagano 1962; Lake et al. 1968; Tingari 1973; Okamura and Nishiyama 1976; Gunawardana and Scott 1977) and the budgerigar (Humphreys 1975a), and is in both these species as in the swan and drake (Maretta 1975a; Marchand 1977) divided irregularly in its length. So far no definite statement can be made about the origin of this material, as Okamura and Nishiyama (1976) confirm in the case of the cock, while Tingari (1973) speaks of a genesis from the cytoplasm. This formation is missing in most other birds including the pigeon, the spermatozoa of which are in many other respects similar (Mattei et al. 1972; Yasuzumi and Yamaguchi 1977). It has, however, though with considerably different shaping, been developed in another set of animal classes independent of the zoological system, for example, in some amphibians (Burgos and Fawcett 1956; Sandoz 1970; Picheral 1972), in the river lamprey (Stanley 1967), and also in a large number of invertebrates, such as crayfish, mussels, sea-urchins, and ticks (Breucker and Horstmann 1968, 1972). Because of its possible participation in the acrosome reaction during fertilization, which has repeatedly been demonstrated in lower animals (Dan and Hagiwara 1967; Nicander 1970), the rod is mostly described as a perforatorium. So far there have been no studies of the acrosome reaction in birds.

The condensation of the nucleus is one of the most striking structural changes during spermatid differentiation. The form of package of the chromatin in a shape capable of transport is species-specific. It shows, however, an enormous multiplicity of variation which is independent of the zoological system (Horstmann 1970). The condensation form in the swan is of the granular type, i.e., the nucleus of the mature spermatid reaches its almost homogeneous appearance by the condensation of granules which grow gradually larger and which close up more and more. The condensation of the nucleus occurs in the same way in the drake (Marchand 1977), in the cock (Nagano 1959, 1962; Nicander 1970; Okamura and Nishiyama 1976; Gunawardana and Scott 1977), in the budgerigar (Humphreys 1975a), and in pigeons (Mattei et al. 1972; Yasuzumi and Yamaguchi 1977), whereas in other birds a nuclear condensation rather of the fibrous type has been observed (Yasuzumi 1956; Yasuzumi and Sugioka 1971). But also the spermatid nuclei of man (Horstmann 1961; 1970), cat (Burgos and Fawcett 1955), some lower vertebrates (Burgos and Fawcett 1956; Nicander 1970), and numerous invertebrates condense in the granular way. Noticeable in the swan is that almost until the end of condensation a thin space underneath the nuclear membrane remains free of granules, a fact which Okamura and Nishiyama (1976) also found to be true in the nuclear condensation of the cock.

In connection with the condensation processes in the nucleus, the appearance of microtubules in the cytoplasm accompanying the nucleus is of special interest. This formation, temporarily present during the differentiation of spermatids, is often termed "manchette" and has been described in numerous species of the animal kingdom and also in several plants (Dustin 1978). Such microtubular systems, although occasionally somewhat varied, have repeatedly been confirmed in birds; they were described and discussed particularly fully in the cock by McIntosh and Porter (1967). With regard to the succession of two different orientations of the microtubules as well as their temporal relationship to the condensation of the nucleus, the results in the cock correspond to those in the swan. The following observations in the cock, however, differ: (1) there are lateral connections between the microtubules of the circular manchette which cannot be observed in the swan in spite of the very regular arrangement of the tubules, and (2) the microtubules oriented parallel to the longitudinal axis of the nucleus do not appear until the nucleus is fully elongated and, moreover, possess a thicker wall than the first microtubules running horizontally around the nucleus. From this the authors conclude that there are two separate generations of tubules, one appearing after the other. The alternative to this suggestion would be the change of orientation of one single tubular generation in the course of the elongation of the nucleus, as Yasuzumi and Yamaguchi (1977) assumed for the domestic pigeon. Which of these two possibilities is true for the swan cannot be stated unequivocally. In any case the process must take place very quickly as transitional arrangements in between the two systems are rarely observed. Intermediate stages of disorientated tubules which only appear before the nucleus has reached its final length make it more probable that it is a case of reorientation of one single generation of microtubules. Okamura and Nishiyama (1976), in their study of special problems of spermatid differentiation in the cock, also think this question to be unresolved.

In the same way there are several ideas about the origin of the microtubules. Whereas McIntosh and Porter (1967) for the cock, and Mattei and his colleagues (1972) for the collared turtle-dove, are of the opinion that the microtubular manchette originates from a collection of uniform tubules in the area of the distal centriole, Yasuzumi and Sugioka (1971) speak of a participation of the plasma membrane in the formation of the microtubules in the love-bird. Finally, in the domestic pigeon Yasuzumi and Yamaguchi (1977) take the nuclear membrane and the amorphous substance adjacent to it as being the initial material. In the swan this question must remain open, since as in the drake (Marchand 1977) there are no unobjectionable criteria to support any of the hypotheses.

The role of the microtubules is seen by many authors as influencing or even supporting the nuclear shape (Baccetti and Afzelius 1976; Dustin 1978). A direct relation between the building up or breaking down of a microtubular manchette and the change in shape of the nucleus of the spermatid was described in numerous species of invertebrates (e.g., Kessel 1967, 1970; Wilkinson et al. 1974) and also of vertebrates (e.g., Burgos and Fawcett 1956; Clark 1967; de Kretser 1969; Rattner and Brinkley 1972). Such a connection has been supposed for various birds. Thus McIntosh and Porter (1967) advocated the theory that the circular manchette would bring about the elongation of the nucleus, whereas the second system of longitudinally positioned tubules might be responsible for the slight curvature of the mature spermatid. The findings of Okamura and Nishiyama (1976) in the cock also seem to strengthen this opinion, as when the manchette is missing one is confronted with irregularly shaped nuclei. Yasuzumi and Yamaguchi (1977) speak of a cyto-skeleton of microtubules which accompanies the elongation of the nucleus in the pigeon.

The idea of the microtubules only having a purely mechanical skeletal function for the nucleus is, however, contradicted by different authors. Examples are cited of animals in which the change of shape of the spermatid nuclei takes place without the participation of microtubules at all, e.g., in a scorpion (Phillips 1974), or without relation to the manchette of microtubules where this is present, e.g., in various marsupials (Phillips 1970; Rattner 1972). Fawcett and his colleagues (1971), who examined the factors possibly influencing the change of the nuclear shape in the spermatids of several mammals, birds, and insects, came to the conclusion that in the pigeon and similarly in the cock, contrary to the opinion of McIntosh and Porter (1967), the mechanical strength in the microtubules is not sufficient to bring about the elongation of the nucleus and the later curvature. The findings in the spermatids of the finch, in which the nucleus is developed as a corkscrew-shaped form, independent of a temporarily present manchette of microtubules, also speak against a purely mechanical function. What is thought to be more probable is the inductive influence of the microtubules on the nuclear condensation, which in turn influences the change in shape of the nucleus. Lanzavecchia and Lamia Donin (1972) interpreted similarly their observations made during the differentiation of spermatids in the earthworm, in which the microtubules do not bring about the elongation of the nucleus by mechanical compression but rather stimulate the beginning of chromatin condensation in the periphery of the nucleus through the exchange of substances.

The propelling apparatus of the spermatozoa in the swan is very similar in structure and development to that in the drake (Humphreys 1972; Maretta 1975b, 1977; Marchand 1977), and also to that in the cock (Nagano 1959, 1962; Tingari 1973; Okamura and Nishiyama 1976; Maretta 1977). At the beginning of spermatid differentiation the formation starts from the two T-positioned centrioles in the area of the Golgi apparatus, as in other animal classes. The further development varies, however, in various points and is simpler than, e.g., in mammals (Fawcett and Phillips 1969; Holstein and Wartenberg 1976). Thus no marked neck piece is developed in ducks, geese, and gallinaceous birds, because the mitochondria of the middle piece join directly to the distal end of the nucleus and therefore leave no room for special structures except for the proximal centriole. Its contact with the nucleus is achieved through five little arms with button-shaped ends, in the swan as in the duck and the cock (Maretta 1977).

In these species the unusually long distal centriole is also particularly striking. In the mature spermatozoon of the drake this reaches a length of up to 2 μm (Maretta 1975b), in that of the cock up to 1.5 μm (Okamura and Nishiyama 1976), and in that of the swan about 1.2 μm. As the distal centriole persists, the flagellum proper with the typical 9 + 2 axoneme here begins just toward the end of the middle piece, contrary to the case in, for instance, mammalian spermatozoa (Fawcett and Phillips 1969; Holstein and Wartenberg 1976); in the latter the distal centriole only represents the initial structure, out of which the axoneme develops directly adjacent to the proximal centriole. The central tubules which Lake et al. (1968) have described in the distal centriole of the spermatozoa in the cock could not be confirmed by Nicander (1970) and Tingari (1973); Maretta (1977) saw them as an exceptional observation and neither are they to be found in the swan. Likewise in the drake (Humphreys 1972), the tubules do not begin before the distal third of the centriole. In the pigeon (Mattei et al. 1972) the distal centriole also persists, but it stays about the same length as the proximal centriole.

Further typical structures of the flagellum are the annulus at the end of the definitive middle piece and the ring-shaped sheath of the principal piece. The annulus is not a formation confined to mammals, as Baccetti and Afzelius (1976) have stated, but is also present in a series of birds. In the swan it is made up of only one component and is relatively weakly developed, and this seems to be true likewise of the cock and the drake (Okamura and Nishiyama 1976; Marchand 1977; Maretta 1977), as well as of the pigeon (Mattei et al. 1972). A participation of the chromatoid body in the development of the annulus, as Fawcett (1972) demonstrated for the guinea-pig, is therefore less probable in these birds. Moreover, the annulus appears in the swan already in its definitive form when in the cytoplasm there are regularly still one or two chromatoid bodies some distance from the implantation point of the proximal centriole. Lake et al. (1968), like Tingari (1973), believe that the annulus in the cock is only a thickening of the cell membrane, which in spermatozoa from ejaculation or from the epididymis is mostly regressed. In the swan the annulus keeps its original size even after spermiation, at least as long as the spermatozoa stay in the lumen of the seminiferous tubules. The ring-shaped sheath which surrounds the axoneme in the principal piece is developed in the same way in ducks and geese, as well as in gallinaceous birds (Nagano 1962; Lake et al. 1968; Okamura and Nishiyama 1976; Maretta 1975b, 1977). It is composed of a uniform fibrous to amorphous material. It shows no separation into single fibers associated with the tubules of the axoneme as is found in mammals, for instance (Fawcett 1965), and is less electron-dense. Marchand (1977), however, describes such a separation into nine strings in the Muscovy duck. A diagram shows that this partitioning can be seen only in the distal section of the principal piece and therefore in the swan possibly corresponds to the irregular decomposition of the ring-shaped sheath in a distal direction.

In addition to the development of the acrosome, the flagellum, and the nuclear condensation, of the remaining structures in the cytoplasm the Golgi apparatus, the annulate lamellae, and the tubular body still deserve special notice. A remarkable feature in connection with the development of the flagellum is the continuous relationship of the centrioles to the Golgi apparatus until the state of late spermatids is reached. The close proximity of both these cell organelles is also typical of the cock; therefore Okamura and Nishiyama (1976) supposed that the plasmalemma of the flagellum developed from the Golgi apparatus. The annulate lamellae can be observed in the swan during almost the whole process of spermatid differentiation and they are thus, contrary to the case in the drake (Marchand 1977), not merely a characteristic of the secondary spermatocytes. What is striking in the swan is the increase in the number of the lamellar layers once they no longer have contact with the nucleus. But the subsequent dissolution of the regular pattern of layers seems to indicate that this phenomenon has no special meaning. A structure which in spermatids has so far been described only sporadically is the tubular body, which can, however, show very different structural aspects throughout the animal kingdom. In the same configuration as in the swan it has only been observed in the domestic pigeon (Yasuzumi and Yamaguchi 1977), the Japanese fresh water turtle (Yasuzumi and Yasuda 1968), and an American lizard (Clark 1967), however, there are differences between species or even classes. Thus the diameter of the tubules seems to vary even if one takes differing fixing substances into account (40 nm in the lizard compared with 16 nm in the swan and 20 nm in the pigeon), or ribosomes can appear in contact with the tubules (pigeons: Yasuzumi and Yamaguchi 1977). Tubular bodies as they have been described in some invertebrates (Folliot and Maillet 1965; Starke and Nolte 1970; Stang-Voss 1972) have a deviating morphological aspect. By a special arrangement of the tubules a picture of so-called undulating tubular bodies arises with contrasting double membranes at the periphery (Starke and Nolte 1970; Stang-Voss 1972). Apart from this there are always several present, contrary to the single body per cell in the swan. All these formations have in common their contact with the smooth endoplasmic reticulum. In the swan there occurs the additional peculiarity that the tubular body in later spermatids comes in direct contact with the annulate lamellae, which are also marked by their contact with the endoplasmic reticulum. This finding has so far not been described in spermatids. There has been just as much speculation about the origin of the tubular body as about its function. In the previously mentioned reptiles it is said to arise out of the Golgi apparatus and then to take part in the development of the acrosome (Yasuzumi and Yasuda 1968), or in the development of the microtubular manchette around the nucleus (Clark 1967). Such a connection does not exist in the pigeon (Yasuzumi and Yamaguchi 1977); because of the associated ribosomes in this case a participation in the exchange of protein materials is thought to be more probable. In the case of the undulating tubular bodies of the invertebrates, the accumulation of proteid and lipid structures (Starke 1971) or the aggregation of degenerating membranes (Stang-Voss 1972) is also discussed.

The electron microscopic examination of the spermatid differentiation in the swan presents a further example of the fact that in birds a detailed specification of spermatozoa exists according to different systematic groups. It is already known from older extensive observations made with the light microscope (Ballowitz 1888; Retzius 1909), and from ultrastructural studies (Humphreys 1972), that in this animal class two large groups of principally different spermatozoa can be distinguished, namely those of the Passeriformes (song-birds) and those of the non-Passeriformes; of the latter indeed species from only a few families have been examined. The spermatozoa of the song-birds are provided with a twisted, spiral-shaped acrosome without a perforatorium, a proximal centriole only, and a sort of undulating membrane (Furieri 1962; Nicander 1970; Humphreys 1972). The latter suggests a similarity to the spermatozoa of amphibians, but as far as their fine structure is concerned they are not homologous. The spermatozoa of the non-Passeriformes, to which the gallinaceous birds, the ducks and geese (and therefore also the swan), the pigeons, and the parakeets belong, correspond mainly to the basic type of a spermatozoon and show similarities to those of the reptiles. Therefore these spermatozoa are said to be of the "sauropsidian type" (Humphreys 1972). If one compares single structures, e.g., the acrosome, the large group can be further split up according to the specifications of different families, as is demonstrated in the pigeons (Mattei et al. 1972).

The Sertoli cells, the lamina propria of the seminiferous tubules, and the interstitial tissue as somatic parts of the testis have been examined by means of electron microscopy in few avian species so far. In those studies the elements engaged in the endocrine system of the male gonad, the Leydig cells, and the Sertoli cells, were of special interest, particularly the relation of their ultrastructure and differentiation to biochemical findings. The Leydig cells in seasonal breeders are subjected to an annual cycle which shows a phase displacement against the spermatogenesis. This has already been presented for the Peking duck, a domesticated breed of the northern mallard (Garnier et al. 1973), as well as for the mute swan (Baumgarten and Holstein 1974), and is therefore not dealt with again here. Unlike the Leydig cells, which show themselves unequivocally to be hormone producers, the Sertoli cells demonstrate a much wider spectrum of functions. That they obviously also take part in the hormone production of the testis has been shown in studies of mammals (Christensen and Mason 1965) and reptiles (Lofts and Choy 1971; Lofts 1972), in which after separation of interstitial tissue from seminiferous tubules in both fractions a biosynthesis of androgens could be found. In the cock too it was possible to demonstrate a synthesizing of androgens in the seminiferous tubules, apart from that in the Leydig cells, by using histochemical and fluorescent microscopic methods (Woods and Domm 1966). It most probably takes place in the Sertoli cells, as is primarily to be deduced from their ultrastructure (Lofts 1972). It is known from numerous examinations of steroid-producing cells in the adrenal gland, the testis or ovary (e.g., Christensen and Fawcett 1966) that in these cells a well developed agranular endoplasmic reticulum, microtubules, and a Golgi apparatus, as well as mitochondria with a tubular inner structure, are typical. With regard to such a correlation Sertoli cells have been studied in recent years, predominantly in mammals (Brökelmann 1963; Black and Christensen 1969; Schulze 1974) but also in some birds: the quail (Lofts 1972; Brown et al. 1975), the cock (Cooksey and Rothwell 1973), and the duck (Garnier et al. 1973; Marchand 1973).

In the swan too the Sertoli cells are characterized by an abundant agranular endoplasmic reticulum. But their mitochondria contain single cristae in their matrix instead of tubules. This is also true of the cock (Cooksey and Rothwell 1973) and the quail (Brown et al. 1975), whereas for the drake there are different findings: Garnier and his colleagues (1973) found in the Peking duck mostly "classic" mitochondria and only very seldom those which contained tubules, whereas Marchand (1973) in the Sertoli cells of the active testis of the Muscovy duck demonstrated mitochondria with an inner vesicular-lamellar structure. Lofts (1972) also speaks of finding such mitochondria in the Japanese quail.

The Golgi apparatus in the Sertoli cells of the swan is not very marked, contrary to the findings in the cock (Cooksey and Rothwell 1973) and the duck (Garnier et al.

1973). Also the nucleus does not exhibit those special features of an increased meta-
bolic activity which it shows in many species of reptiles (Dufaure 1971), mammals
(Brökelmann 1963; Schulze 1974), and birds (Cooksey and Rothwell 1973; Marchand
1973). The nucleus is always round to oval in shape and has no indentations. Para-
plasmic inclusions such as glycogen and lipid droplets are almost always absent from
the Sertoli cells of the swan; in various species they appear as characteristic (Dufaure
1971; Schulze 1974) and possibly influence the process of spermatogenesis (Lacy
1962). In this case the ultrastructure gives no certain evidence on hormone activity,
although one can hardly believe that the Sertoli cells in the swan function fundamen-
tally differently from those of other vertebrates.

Microtubules and filaments are regular components of the Sertoli cells in the swan,
but they seldom show a favored direction of orientation such as Christensen (1965)
found, for example, for the microtubules of the Sertoli cells in the guinea-pig. Nevertheless it is possible that in the swan too they may have something to do with the continuous cellular movements in the germinal epithelium and may take part in these, or
they could serve as a sort of cyto-skeleton for the extended Sertoli cells, a possibility
which was also discussed for the cock (Cooksey and Rothwell 1973). The same goes
for the filaments, sometimes arranged in bundles, which as a rule lie in the inner part
of the cell with no connection to the cell membrane; special junctional complexes between neighboring Sertoli cells, above all in the basal area of the germinal epithelium,
to which the filaments could be connected, are absent in the swan. Such contact specializations have many times been shown to be typical for the maintenance of the
blood-testis barrier between the basal and adluminal compartments of the seminiferous tubules in mammals (Flickinger and Fawcett 1967; Nicander 1967; Dym and
Fawcett 1970; Bigliardi and Vegni Talluri 1976; Dym and Cavicchia 1977; Connell
1978; Russell 1978). In birds the question of the location of such a barrier still remains
open. Possibly a tracing with Lanthanum during the spermatogenically active phase of
the swan in spring would show whether there are any structures responsible for this
separation, and if so, which. In the cock, Cooksey and Rothwell (1973) described
special Sertoli cell – Sertoli cell contacts, but without saying whether they were predominantly present in the basal germ cell compartment, whereas Brown and his
colleagues (1975) found Sertoli cell contacts in the quail in the form of depositions
of electron-dense material, mainly near the lumen. Such cell contacts in the apical cell
region can also be observed in the swan, but only during the short period of time when
a real lumen is present in the seminiferous tubule.

On the other hand, in the swan contacts more frequently exist between Sertoli
cells and germ cells, mostly spermatogonia and early primary spermatocytes. These
contacts consist only of electron-dense material positioned symmetrically to the
parallel running cell membranes from inside the cells. Similar connections have also
been observed in man (Altorfer et al. 1974), more frequently in fetal testicular tissue.
Kaya and Harrison (1976) and Russell (1977a) also described contacts between Sertoli
cells and early germ cells in the seminiferous tubules of the adult rat, but there they
found additional filaments on the side of the Sertoli cells. They were interpreted to be
places of exchange for nutritional substances or taken as being responsible for the
mechanical stiffening of the Sertoli cells during the moving upward of the spermatocytes from the basal to the adluminal compartment.

Much more widespread are contact specializations of most different kinds between
Sertoli cells and later developmental stages of the germ cells. In the same way, these

are thought to have a mechanical function before and during spermatid differentiation (Russell 1977b), so that the final loosening of the contacts causes the spermiation (Ross 1976). These findings in the rat and mouse can be established in a similar way in the cock (Cooksey and Rothwell 1973), but again not in the swan; here the contact is confined to the encircling of the almost cytoplasm-free acrosome by cytoplasmic processes of the Sertoli cells, without specially dense areas or substructures, as they were described in man (Horstmann 1961) or in the hamster (Vitale-Calpe and Burgos 1970). Altogether there are no noticeable changes in the apical region of the Sertoli cells during spermatid differentiation in the swan, so that the elements which might be responsible for the release of the spermatids into the tubular lumen cannot be determined ultrastructurally.

In connection with the spermiation, the Sertoli cells have to take over another function, namely the phagocytosis of residual bodies. The disposal of the left-over cytoplasmic components of the spermatids from the germinal epithelium is especially marked in gallinaceous birds and ducks, and so in the swan; cytoplasmic droplets in the freely moving spermatozoa are absent (Tingari 1973; Marchand 1977), in mammals on the other hand they are typical. That the Sertoli cells are capable of phagocytosis has been known since the examinations made by Rolshoven (1947–1948) and Roosen-Runge (1955), and has been proved by injecting foreign particles, e.g., dyes or suspensions of carbon, into the seminiferous tubules (Clegg and MacMillan 1965; Carr et al. 1968). Such foreign particles are indeed only taken up by the Sertoli cells, whilst incorporated residual bodies are extensively broken down. The phagocytosis of residual bodies is described in most different species of higher vertebrates, and frequently in this process lipid droplets are accumulated – whether they are indigestible remains or new formations is uncertain (Lacy 1960, 1962; Brökelmann 1963; Sapsford et al. 1969; Dufaure 1971; Fouquet 1974; Humphreys 1975b). In the Sertoli cells of the swan the breaking-down stages of the residual bodies are recognizable in the form of pleomorphous osmiophilic inclusion bodies. There are, however, no signs that lipids either remain as end-products or are newly synthesized during the degradation of the residual bodies. That the Sertoli cells phagocytose not only residual bodies, but degenerating germ cells too, can also be observed in the swan. The increased destruction of degenerating germ cells is, however, the main feature of the involution phase in the annual cycle of spermatogenesis, and will be discussed under this particular aspect.

The lamina propria of the seminiferous tubules in the swan exhibits the fundamental organization of fibroblasts, myofibroblasts and fibrous material in alternate layers, which is well known in mammals (Dym and Fawcett 1970; Bustos-Obregón and Holstein 1973) and which in birds has hitherto only been described in the cock (Rothwell and Tingari 1973). Similar to the cock, a characteristic also of the swan is that the first layer of cells adjacent to the periphery of the tubule consists of real fibroblasts; only the following layers contain the characteristic structures known for the myofibroblasts.

That the lamina propria is well supplied with unmyelinated nerve fibers is known not only in the cock (Tingari and Lake 1972b) but also in the swan, and has already been extensively described (Baumgarten and Holstein 1968, 1974).

7 The Annual Cycle of the Male Gonads (Light Microscopic Studies)

The confinement of reproduction to an ecologically favorable period of time leads in the swan to a marked annual cycle. Apart from the obvious phases of activity in the female, the production of male germ cells capable of fertilization is also confined only to a short duration of time within a year, usually to the months of April and May in temperate northern latitudes. The preliminaries for the attainment of this goal begin already at the end of January/beginning of February. On the other hand the involution of some seminiferous tubules begins at a time when parts of the testis are still fully active. Because of the not entirely uniform development of the organ, but also because of the extensive degradation processes inside the germinal epithelium, the regression phase stretches over a comparatively long period of time, right into autumn. It continues into the resting phase which is followed again by the beginning of a new annual cycle (Figs. 20a, b).

The seasonal building up and breaking down of the testis is already recognizable from the outside by the considerable change in size of the organ. Due to individual variation, the measurement of volume or weight gives only a rough indication, however. Moreover, such measurements are affected not only by the changes in the germinal epithelium but also by those in the lamina propria and in the interstitial tissue (Figs. 20a, b), whose phases do not necessarily run parallel to those of spermatogenesis. A much more decisive criterion for the rate of spermatogenic activity is the increase and decrease in diameter of the single seminiferous tubules within the course of a year. From a value of 70–90 μm during the resting phase from the end of November until the end of January, the diameter increases more than threefold to reach the maximal value in April, and afterwards it quickly decreases again (Table 1). The minimum tubular diameter, however, is only reached again in late autumn, because of the long-lasting extensive involution process.

Although preparations were made and studied every month, the light microscopic appearance will be described in terms of five phases, because the changes in the germinal epithelium during the course of a year are thus shown especially clearly. So far in the literature the single developmental states of the germinal epithelium in the different seasons have been termed "stages". Instead of this the term "phase" will be used, in

Table 1. Graphic representation of the change in diameter of the tubules during a year. The *white area* above the *spotted area* in each column indicates the degree of expansion

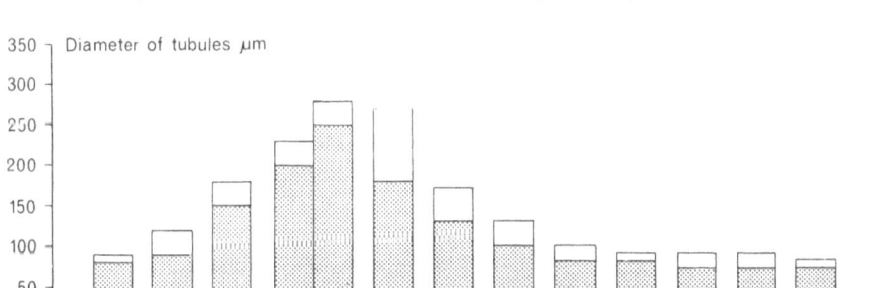

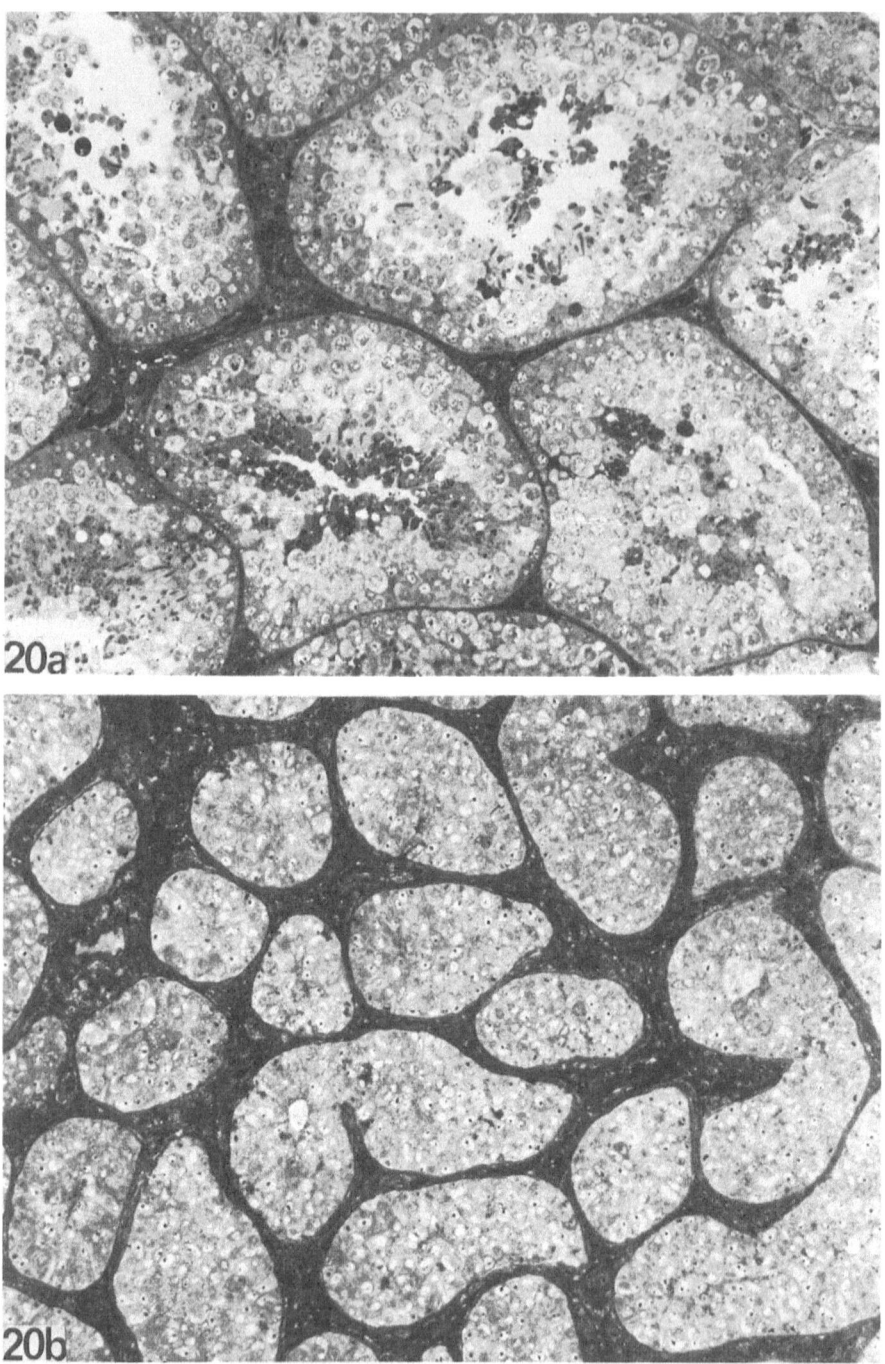

Fig. 20a, b. Histological sections from testes at the time of maximal spermatogenic activity in April a and during the resting phase in December b. Noteworthy are the different dimensions of the seminiferous tubules and the changed proportion of tubules to interstitial tissue. × 200

order to differentiate unequivocally between the course of development inside the annual cycle and the period of maximum activity of the germinal epithelium in spring. The classification of stages which can be undertaken only when the full spermatogenic activity is reached ought not to be confused with the annual processes in the germinal epithelium, although there are certain points of overlapping.

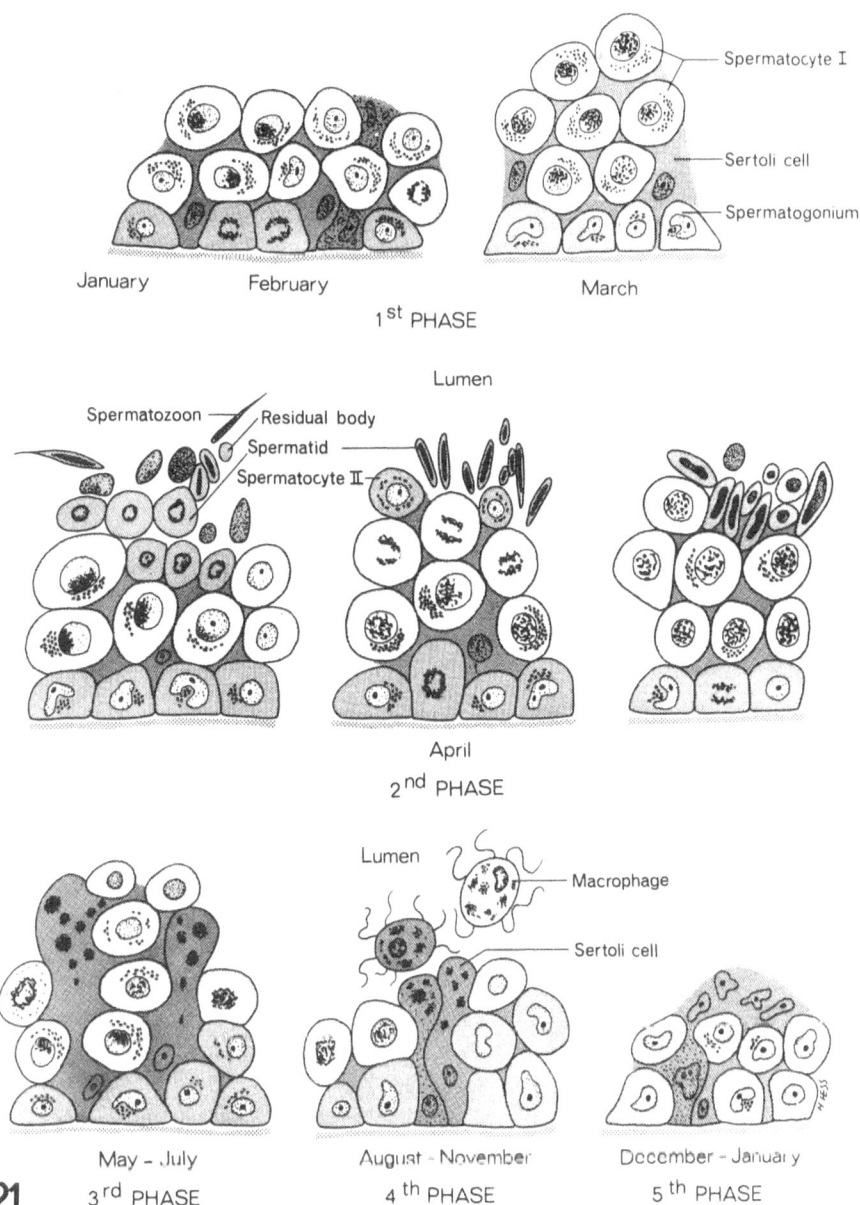

Fig. 21. Schematic representation of five phases of the annual cycle in the swan on the basis of semithin sections

7.1 Phase 1: Multiplication Period

Phase 1 begins at about the end of January/beginning of February and is above all characterized by the multiplication of spermatogonia, which have remained during the winter resting period as stem cells (Fig. 21). The increased mitotic activity occurs in groups and is to be observed over the whole of the transverse section of the tubule as long as the spermatogonia are the only germ cells present. With the growing number of cells, the diameter of the tubule increases during February to 150–180 μm (Table 1). The cells which happen not to be in division show the characteristics of a spermatogonium, namely a round or kidney-shaped nucleus with one or two clear nucleoli and the lighter, homogeneous EDTA-positive body about the size of the nucleolus, as well as the collection of cell organelles at one side of the nucleus. In between these spermatogonia one occasionally finds, singly or in groups, strongly colored cells the nuclei of which, in spite of their density, show them also to be spermatogonia. They are stretched out with several pointed processes and their cytoplasm can contain vacuoles. Details about their fate are unknown, presumably they disintegrate. With further development of the germinal epithelium they suddenly disappear, although the seminiferous tubules have no lumina at this time and the Sertoli cells only seldom show signs of phagocytosis. The nuclei of the Sertoli cells nearly all lie along the basal membrane between the spermatogonia.

With the appearance of primary spermatocytes in interphase or in the first stages of meiotic prophase the spermatogonia are pushed more and more to the periphery of the tubule, where they finally position themselves along the basal membrane in one layer (Fig. 21). In doing so they force the nuclei of the Sertoli cells out of their basal position into the layer of the spreading primary spermatocytes. Then most of the nuclei of the spermatogonia are long and bent or kidney-shaped, which means that the multiplication period, which precedes the building up of the germinal epithelium to the completion of spermatogenesis, has finished. After this, mitotic divisions of the spermatogonia only occur again in combination with particular stages of development of spermatocytes and spermatids, as they were described in the stages of spermatogenesis.

7.2 Phase 2: Complete Process of Spermatogenesis

Phase 2 is characterized by the complete process of spermatogenesis, and therefore need not be represented in detail (Figs. 2, 21). It starts relatively short-dated, i.e., not before it is expected that the weather conditions give promise of a successful mating and subsequent breeding. Therefore the transition from phase 1 to phase 2 can be extended to varying degrees. From the comparison of testicular specimens taken from birds in several different years during the months February to May, it can be supposed that after the multiplication phase the primary spermatocytes can once more interrupt their development for several weeks in interphase, in leptotene, or perhaps also in zygotene. Only once the outside temperature rises and remains for a certain period at 10 °C or more are they stimulated to develop further and quickly complete both meiotic divisions.

With the appearance of differentiating spermatids a development in cell groups becomes clearer and clearer, and these order themselves in particular combinations and finally make it possible for spermatogenesis to be divided into stages. A lumen appears in the seminiferous tubule when the first spermatids are so mature that they will shortly be released from the germinal epithelium (Figs. 20a, 21). At the height of spermatogenic activity there are lumina present in about 50% of the tubules. The tubular diameter then measures $250-280$ μm (Table 1).

There are only very few single degenerating cells in this phase. These are either primary spermatocytes in the first stages of meiotic prophase or spermatogonia. The primary spermatocytes can easily be recognized by their swollen nuclei with clumpy karyoplasm and an almost structureless cytoplasm. The degenerating spermatogonia appear especially dense and strongly colored in comparison. They become detached from the basal membrane and move towards the center of the tubule, recognizable by their very irregular shape.

7.3 Phase 3: Beginning of the Regression Period

In phase 3 the involution of the germinal epithelium begins (Fig. 21). It starts first in only a few tubules but in the course of $6-8$ weeks spreads over the whole organ. Then the diameter of the tubule decreases to $120-160$ μm (Table 1). The regression first becomes noticeable in that once a generation of mature spermatids have been released into the lumen of the tubule, the cell generations which remain in the germinal epithelium more or less inhibit their development, so that finally an association with stages of spermatogenesis is no longer possible. Nearly everywhere the lumen in the tubules disappears; the differentiation of spermatids does not go further than the primary stage of the round nucleated cell. By far the largest area of the germinal epithelium is made up of primary spermatocytes at the beginning of meiotic prophase. Their synchronized development in groups is frequently given up, instead of which single cells at different developmental stages lie next to each other. The spermatogonia, mostly with round nuclei, in which mitosis can occasionally be observed, are predominantly positioned in a single layer along the basal lamina.

Several times, however, this layer is interrupted by nuclei of Sertoli cells, which are relatively numerous in this phase. Additionally to this position one also finds Sertoli cell nuclei distributed over the whole of the tubular transverse section, occasionally right up to the center of the tubule. The cytoplasm of these cells is often darker than that of the germ cells and is therefore easy to recognize in its fairly branched form. In the supranuclear region the cells contain many very dark inclusions which increase in size and amount towards the center of the tubule. This obvious increase in activity of the Sertoli cells, since it is never seen in such a high degree at other times of the annual cycle, points to an increased disintegration of germ cells which are no longer needed that season.

Above all zygotene and pachytene seem to be particularly sensitive stages during which primary spermatocytes degenerate singly or sometimes in groups. Electron micrographs support this, in that the synaptonemal complexes of paired homologous chromosomes can be observed very frequently and also for an especially long period in the nuclei of highly degenerated cells (Fig. 22). Under the light microscope the nuclei appear to be pycnotic or swollen with peripherally located chromatin and are surrounded by a very dense cytoplasm. Spermatogonia very seldom degenerate at this time, except in some tubules where they degenerate in groups. They then lie, with their strongly colored and often vacuolized cytoplasm, together with a deformed nucleus pressed flat against the basal membrane. From there they can sometimes reach like a septum to the center of the tubule, where they may perhaps be expelled.

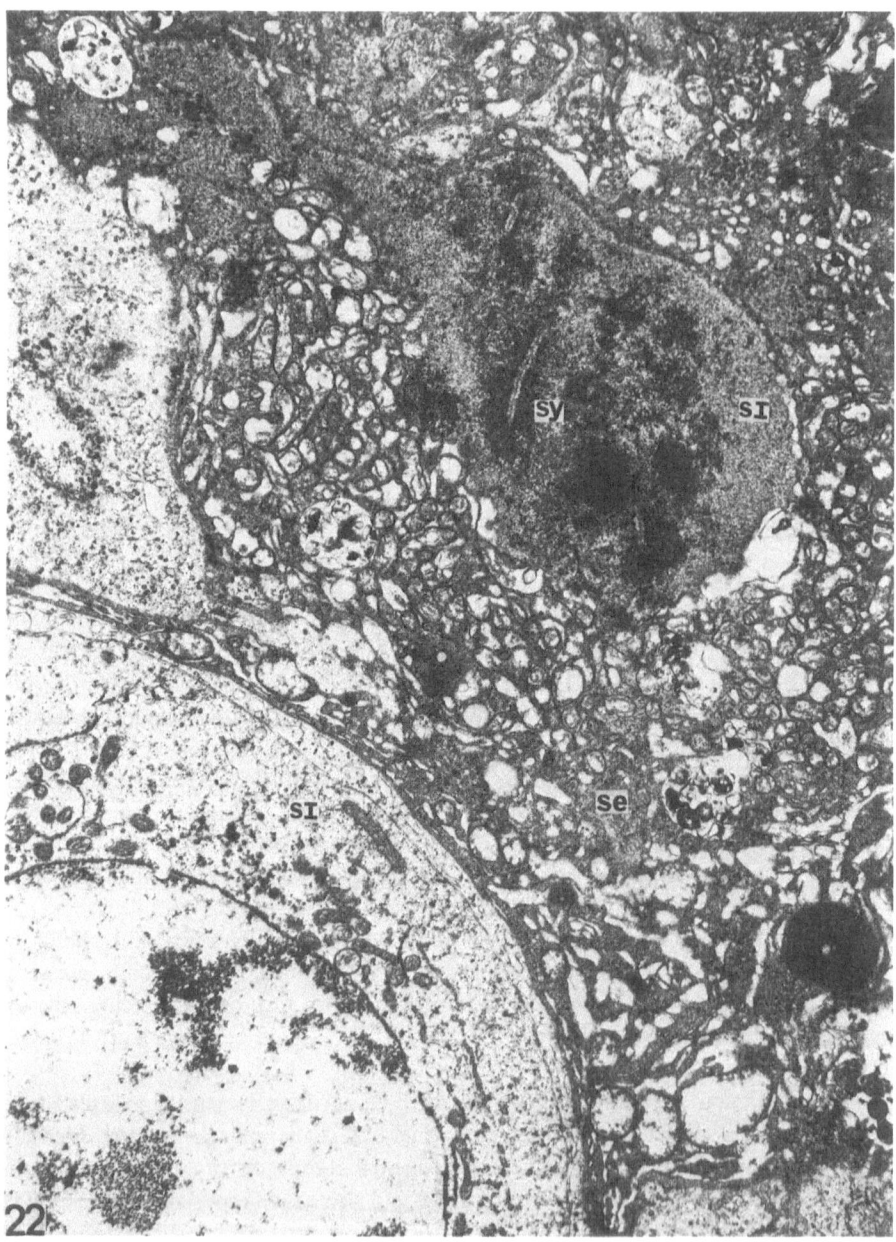

Fig. 22. Section from a seminiferous tubule during phase 3, at the end of July. The remains of a degenerating primary spermatocyte from the zygotene still show the resistant synaptonemal complexes. *se*, Sertoli cell; *sl*, primary spermatocyte; *sy*, synaptonemal complex. × 9200

7.4 Phase 4: Invasion of Macrophages

Phase 4 has to be regarded as a special period of the regression phase (Fig. 21). It has only a relatively short time-span and sometimes escapes observation because of this, and also because of large individual variations. Possibly in some cases it appears more than once during the involution phase of one year, since it could be observed in August and also in a diminished form in November. This phase is characterized by the new appearance of lumina in a series of tubules (Figs. 23a, b). But above all one finds in these lumina as well as in the center of closed tubules numerous round cells, at the periphery of which extremely thin processes can be recognized, especially when they are lying free. These cells are laden with a great amount of very dark inclusions, so that one can often hardly make out their small nucleus with clear nucleolus and peripherally positioned chromatin (Figs. 21, 23a, b). These are macrophages which help the Sertoli cells in the disposal of degenerated cells when the breaking down of the germ cells is at its height. Probably most of them are transported over the tubular lumen out of the seminiferous tubules. One can differentiate them from Sertoli cells by their ultrastructure alone, and then follow their way from the interstitial tissue through the germinal epithelium.

In this phase of the invasion of macrophages, apart from sporadic groups of primary spermatocytes in leptotene the germinal epithelium consists predominantly of spermatogonia, which often lie in two or three rows one upon another. They have their characteristic appearance, with round or kidney-shaped nuclei on one side of which the cell organelles are concentrated. Frequently, the basal cells are somewhat elongated, whereby the nuclei are positioned with their longitudinal axis vertical to the basal lamina (Figs. 21, 23b). Thus obviously a closer packing of the cells is possible, for in the meantime the diameter of the tubules has decreased to $90-110\,\mu m$ (Table 1).

The nuclei of the Sertoli cells at this time occur singly or in groups of up to four, nearly all positioned close to the basal lamina between the spermatogonia, whereas their cytoplasm can be followed as thin, darkly colored processes often upward to the lumen or to the center of the tubule (Figs. 21, 23b). Also here the dark inclusions of the cytoplasm in a central direction increase in number and size. In those tubules in which a lumen is present the apex of the Sertoli cells, which is filled with inclusions and dark droplets, frequently appears to be bulged balloon-like into the lumen.

7.5 Phase 5: Winter Resting Period

Phase 5 represents the end of the annual cycle and is a marked resting phase (Figs. 21, 23c, d). The tubules have a diameter of only $70-90\,\mu m$ (Table 1), a lumen is seldom and only temporarily present. The spermatogonia, which are the only germ cells present, are mostly positioned in two layers. Their cytoplasm contains numerous cell organelles which are evenly distributed. Besides some round nuclei, there are predominantly irregularly shaped longish nuclei present, in which the nucleolus and the homogeneous light body can well be recognized. Between the basal spermatogonia lie the nuclei of Sertoli cells — as a rule singly — which at this time frequently contain one particularly prominent nucleolus, or perhaps two. Some of the Sertoli cells exhibit a very dark cytoplasm from base to the apex, so that the cell branches can be seen clearly (Fig. 23c). Near them, either at the tubular base or on the way to the center, are to be found single

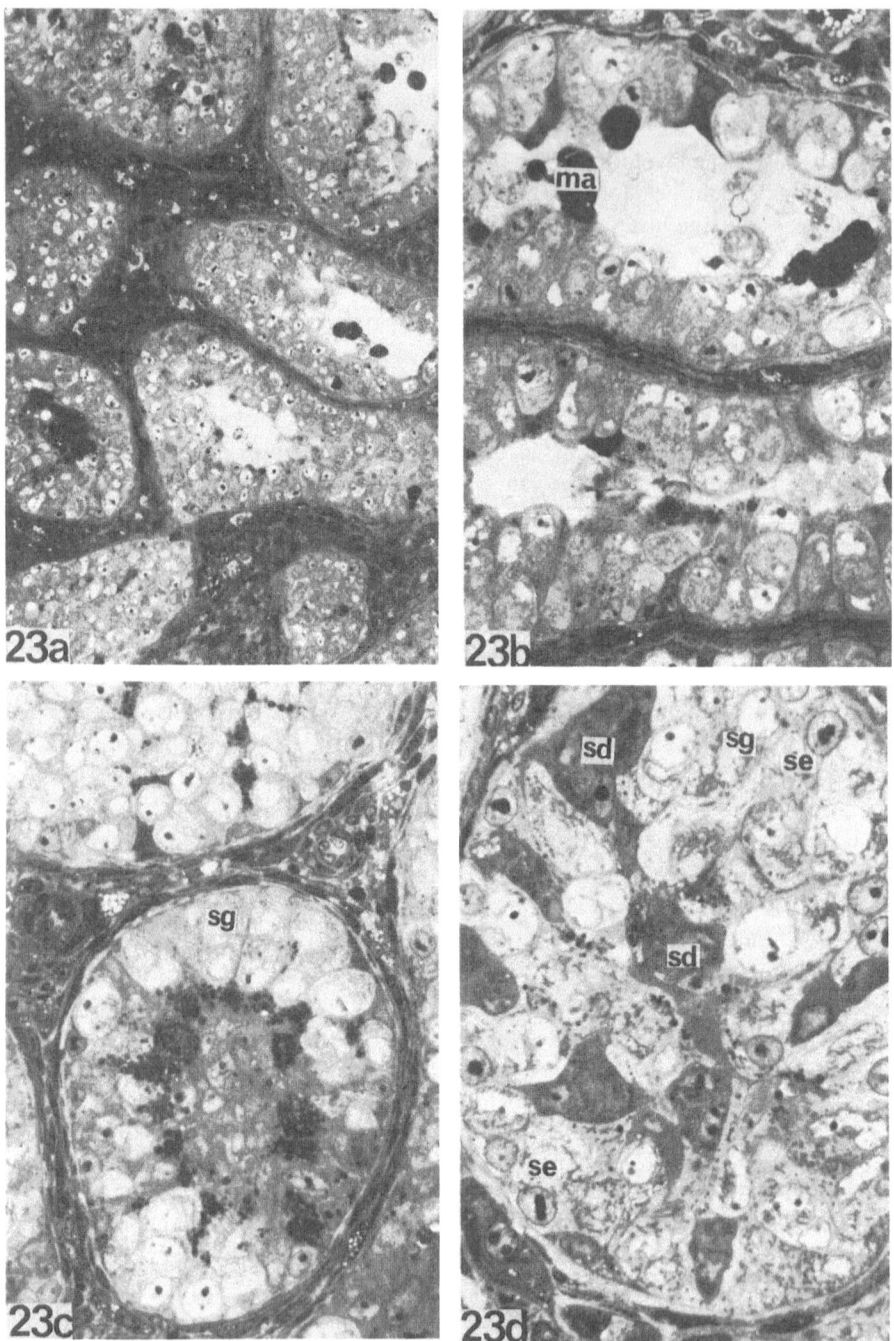

Fig. 23a–d. Light microscopic illustration of phases 4 and 5 of the annual cycle. *a* and *b* At the climax of involution, here at the end of August. *c* End of involution phase at the beginning of November. *d* Resting phase at the middle of December. *ma*, macrophage; *sd*, degenerating spermatogonium; *se*, Sertoli cell; *sg*, spermatogonium. *a* × 225, *b*, *c* × 560, *d* × 900

degenerating spermatogonia, which can still be identified in spite of their strong coloring (Fig. 23d). But usually one cannot see much difference in color between the cytoplasm of the spermatogonia and that of the Sertoli cells, the cytoplasm of which at best contains a small number of dark droplets apically towards the center of the tubule. In some tubules these droplets can be present simultaneously in all Sertoli cells in a greater quantity, so that they seem to flow together and form a ring of dark particles near the center (Fig. 23c). This ring surrounds a homogeneous lighter area in which single polymorphous cell nuclei are found, which probably belong to extensively degenerated spermatogonia. Perhaps these tubules are in the state immediately before the formation of a new lumen, or possibly also before a repeated invasion of macrophages, as was observed in two swans from different years in November.

7.6 Discussion

On the seasonal changes in the reproductive behavior of birds an extraordinarily large number of field observations have been made, mostly over many years, and could often have been performed only with the assistance of amateur ornithologists (e.g., Payne 1969; Hilprecht 1970) or support given by national organizations (e.g., Wildfowl Trust, England; Division of Wildlife Research, Australia). They have built up the basis for studies of the normal reproductive cycles and experiments affecting them. A wealth of publications has thus resulted, mostly in the areas of physiology and endocrinology, less so in that of morphology. (For a review of the literature see Murton and Westwood 1977.)

The increase and decrease in the weight and volume of testes during the reproductive cycle is frequently thought to be sufficient for the evaluation of the building-up and breaking-down processes in the gonads (Wright and Wright 1944; Hiatt and Fisher 1947; Farner and Wilson 1957). Indeed, the trend of development can be stated in this way, but considerable individual variations influence these measurements to such an extent that they can only approximately be related to the stage of development in the testis, as Johnston (1956) demonstrated impressively for the gull and Johnson (1961) for the duck, and as is also true for the swan. Likewise for the duck, Schöneberg (1913) gave an example of an undifferentiated testis being bigger than one in the middle of its building-up stage. Moreover, Blanchard and Erickson (1949), in their examination of the annual cycle of a species of sparrow, pointed out that the first histological changes in the seminiferous tubules and in the interstitial tissue, which signify the start of spermatogenesis, appear several weeks earlier, before an increase in volume of the testis is measurable at all. And furthermore, for the changes in weight and volume, the differing development and regression of tubular and extratubular tissues, which normally take place somewhat out of phase with each other, have to be taken into account (Blanchard and Erickson 1949; Marshall 1961a; Johnson 1966; Baumgarten and Holstein 1974). Therefore measurements of the changes in diameter of the seminiferous tubules such as have been carried out in the swan and other birds, e.g., the arctic fulmar (Marshall 1949a), the rook (Marshall and Coombs 1957), the jackdaw (Threadgold 1956–1957b), different species of blackbirds (Payne 1969), and the tree sparrow (Chan and Lofts 1974), give a much better idea of the stage of development of the germinal epithelium. Even using this method, one finds individual variations of the tubular diameter within a species, as well as from species to species, and these are somewhat more obvious at the time of maximum spermatogenic activity, as is shown in the measurements made in the swan in April. From the winter resting period to the breeding season, the value of the tubular diameter rises by three- (arctic fulmar) to sixfold (sparrow) and ranges in the species so far studied (of very differing taxonomic relationships) from a minimum of 40 μm in winter to a maximum of approximately 300 μm in spring. The swan accords well with these data, having values of approximately 70 μm in winter and 280 μm in April. The values given by Mori and George (1978) for the seasonal changes of the tubular diameter in the Canadian goose lie more than a tenth power lower and therefore appear rather doubtful.

The building up of the germinal epithelium to reach complete spermatogenesis and the following involution to reach an absolute stage of rest is represented histologically in the swan as it is in other birds. Differences concern the duration of single developmental stages as well as the synchronization of the differentiation processes inside the seminiferous tubules which as already mentioned (see p. 17) is much more marked in the song-birds than in the non-Passeriformes. Such a synchronization of the single cell generations over larger areas of the testis simplifies to a great extent the statement of building-up and breaking-down phases of the germinal epithelium as shown in the detailed description of the North American white-crowned sparrow (Blanchard 1941), the starling (Bullough 1942), and two species of blackbird (Payne 1969). Of the birds not belonging to the Passeriformes Johnston (1956) therefore described the annual cycle of the Californian gull with reference to Blanchard (1941), and Johnson (1961) presented that of the duck with reference to Johnston (1956). However, mostly unnoticed, Schöneberg (1913) had already described the seasonal changes in the germinal epithelium in the duck, which are similar to those in the swan.

The division of the reproductive cycle into stages, a term replaced by the word "phases" in the present study, varies somewhat from author to author. Besides the four main phases (resting phase, start of growth, complete process of spermatogenesis, and regression), to which Marchand and Gomot (1973b) confined their description of the annual cycle in the duck, the changes from the resting phase up to the phase of full activity have mostly been further subdivided, e.g., the appearance of each new cell generation or the changes in them were each defined as a particular stage. Such short steps can be isolated more exactly where there is extensively synchronized development than where there are heterogeneous manifestations as found in the swan or duck. Johnson (1961, 1966) has therefore tried to quantify his findings by dividing the process in the duck into six stages, from the winter resting period through first changes in the interstitial tissue up until the completion of spermatogenesis, and determining the percentage of the single cell categories in each stage.

During the building up of the testis special attention should be paid to that particular period which was described in the swan as a transition from the first to the second phase. At this time in the development a temporary standstill can occur, which is recognizable as the unchanging picture of the primary spermatocytes in the early stages of meiotic prophase. This fact has so far been given little notice in other avian species, except for the examination of the annual cycle of two species of sparrows made by Blanchard (1941), although it is probably of fundamental relevance with regard to the effects of exogenous factors on the reproductive cycle. Blanchard stated that, as in the swan, the yearly growth of the gonads nearly always begins in January, while the stage in which primary spermatocytes in synapsis can be observed is found in different years at different points of time depending on the outside temperature. The earlier in a year milder temperatures occur, the earlier the maturation of the germ cells begins. The following period which is necessary for the development from the middle of meiotic prophase to completion of spermatogenesis, is again of very constant duration. Thus in the swan a possible delay of the whole breeding period by 6–8 weeks can result from unfavorable temperatures, i.e., the start of activity can be displaced from the middle of March to the beginning of May, as the very different climates in the years 1973–1978 have shown.

These facts point to the significance of environmental factors which affect the annual reproductive cycle. In the swan, light seems as significant as temperature, at least for initiating spermatogenic activity. Phase 1, the period of multiplication of the spermatogonia, seems to be initiated by the increasing amount of daylight more or less invariably at the end of January. The start of phase 2, however, is dependent mainly on temperature conditions. These interactions of different environmental factors have also been emphasized and discussed by other authors in a series of more physiologically and behaviorally oriented studies (Marshall 1949b, 1961b, 1970; Lofts and Murton 1966; Immelmann 1971; Menaker 1971; Gwinner 1973).

8 The Regression Period (Electron Microscopic Studies)

8.1 Germ Cells

The results of the light microscopic studies on the single phases of an annual cycle can be confirmed by electron microscopic examinations. Since the characteristics of the different types of germ cells, as far as they can be met out of the active phase of spermatogenesis, do not change, it is not necessary to repeat the description of their ultrastructure. However, when these cells disintegrate, then their density often changes so much that certain characteristics can only be recognized and identified by means of electron microscopy (Fig. 22). With the increased appearance of degenerating germ cells, the Sertoli cells also undergo changes such as are only observed during the regression phase of the testis, on account of increased phagocytotic activity. Furthermore, only by examining the ultrastructure is it possible to characterize the transiently appearing macrophages and to trace them on their way from the interstitial tissue across the germinal epithelium. Therefore that phase in the process of involution which was decribed light microscopically as phase 4 of the annual cycle will now be described and discussed in the light of electron microscopy.

In this phase, the germinal epithelium contains mostly spermatogonia and Sertoli cells (Fig. 24). Primary spermatocytes can still appear in small groups, but mostly they develop only to leptotene or early zygotene with the beginning of chromosome conjugation. Premature germ cells are only present at the beginning of the phase. The spermatogonia are often positioned in two or three layers on top of each other and are almost never connected with one another by intercellular bridges. Even in this phase they only touch the basal lamina with a few small feet, regardless of whether their longish nucleus is oriented parallel or perpendicular to the base of the tubule (Figs. 24, 25). Here and there spermatogonia are found near the center of the tubule, and these are very rich in contrast. This possibly indicates the beginning of degeneration.

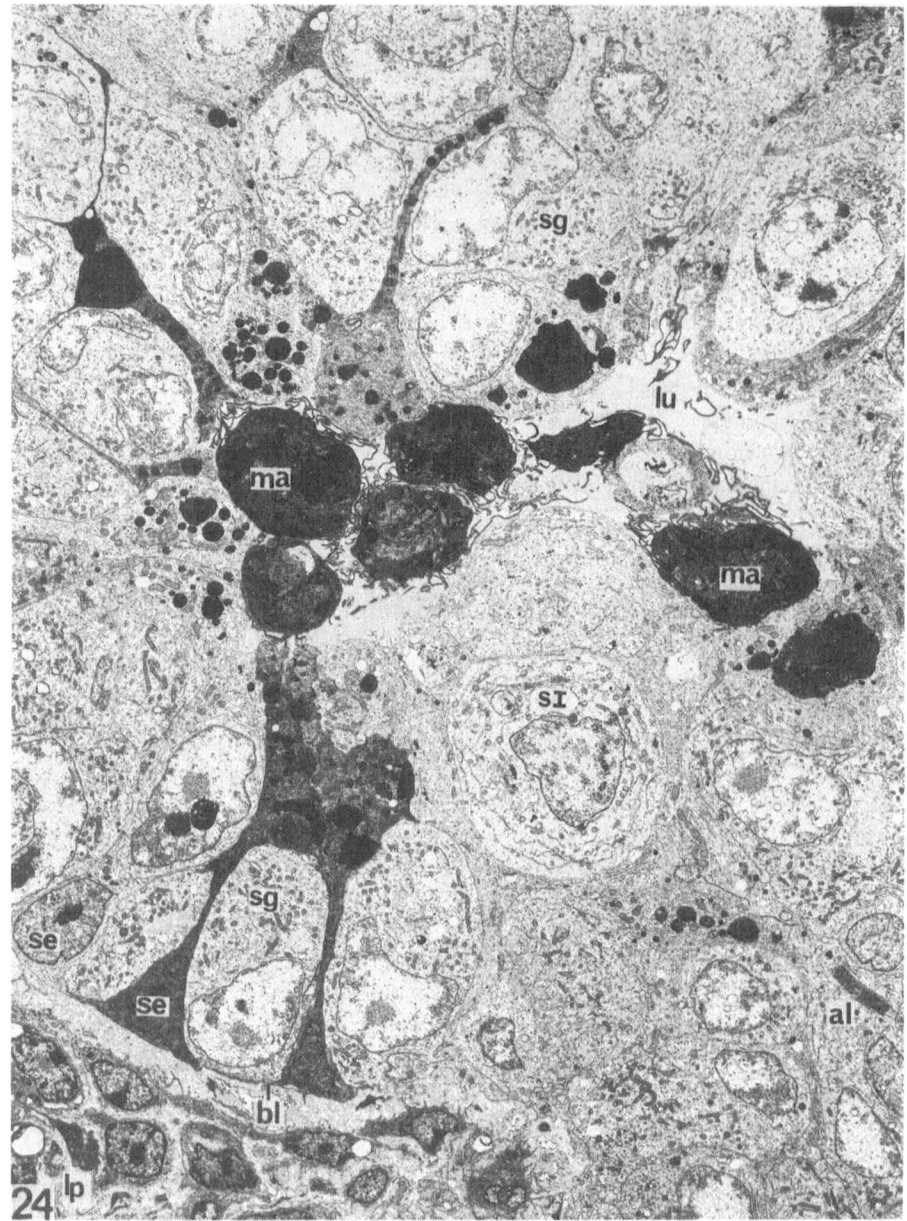

Fig. 24. Survey of a seminiferous tubule during the invasion of macrophages at the end of August. *al*, annulate lamellae; *bl*, basal lamina; *lp*, lamina propria; *lu*, tubular lumen; *ma*, macrophage; *se*, Sertoli cell; *sg*, spermatogonium; *sI*, primary spermatocyte. × 2000

8.2 Sertoli Cells

Between the spermatogonia lie the Sertoli cells, singly or several juxtaposed. Either they are in contact with one another by their long lateral processes with which they surround the germ cells, or they border on each other directly with their lateral cell membranes touching over a larger area (Fig. 24). Special cell contacts are not very frequent in the basal or middle areas of the cell; either they are present as tiny inconspicuous desmosomes, or both cell membranes are covered with electron-dense material at the cytoplasmic side over a larger area, whereby the intercellular space keeps its normal width. But this type of cell contact can occasionally also be observed between Sertoli cells and spermatogonia, as mentioned earlier. What is striking, however, is that in such seminiferous tubules, which contain a lumen in this phase, apically where they partly border the lumen the Sertoli cells are regularly connected to each other through a special strengthening of their membranes (Fig. 25). With the methods of electron microscopy used in the present study the type of cell contacts cannot be more precisely classified, but the intercellular clefts seem to be more or less closed off from the lumen.

The Sertoli cells frequently sit on the basal lamina with broad feet (Figs. 24, 25), whose processes still continue underneath the spermatogonia and lift the latter from the basal lamina except for a few points of contact (Fig. 25). In these broad feet one finds the nuclei of the Sertoli cells, which as a rule lie between the spermatogonia of the basal layer and seldom in the second one. In general the nuclei are oval-shaped or triangular with the same inner structure that they have at the time of complete spermatogenesis. Sometimes, however, they can be seen with one or two deep indentations and an increased number of nuclear pores.

The cytoplasm of the Sertoli cells in this phase of the annual cycle shows a broad spectrum of variations in shape, number, and distribution of its components. This is an expression of increased activity of the cells, which either present themselves in very different functional states or represent different cell types. Common to them all is their richness in endoplasmic reticulum, as at the time of complete spermatogenesis (Fig. 25). In the lighter appearing cells it is present predominantly in the smooth form; the vesicles and irregularly shaped cisterns are loosely distributed over the whole of the cytoplasm (Fig. 26b). Only now and then are ribosomes attached to the profiles of the endoplasmic reticulum; usually they appear singly or in moderate numbers as polysomes. Moreover, roundish bodies sporadically occur, surrounded by a membrane with homogeneous electron-dense contents having a diameter of 0.2 to 0.3 μm (Fig. 25).

In the darker cells the endoplasmic reticulum is considerably augmented (Fig. 24). Simultaneously there is a more dense packing of the cisterns and tubules, to which larger numbers of ribosomes can attach themselves. With increasing ribosome association the profiles of the endoplasmic reticulum become partly parallel oriented. Another striking feature of these cells is the occurrence of the numerous, often very large inclusion bodies, which can reach a diameter up to 5 μm (Figs. 24, 26a). Obviously they are secondary lysosomes, as they are surrounded by a membrane and show an extremely heterogeneous contents in the form of differently structured subunits probably resulting from a preceding phagocytosis. These inclusions are almost without exception found supranuclearly and often they appear in greater number in the cell apex, where they bulge balloon-like into the lumen of the tubule (Fig. 24). The increase in density of the cytoplasmic matrix frequently but not necessarily conforms to the growing number of inclusions. After heightened phagocytotic activity the tubules of the endoplasmic reticulum can often be very widened (Fig. 26a, 30a), as a result of which an additional densifying of the cytoplasm occurs, and therefore these cells are much more contrasted than the germ cells.

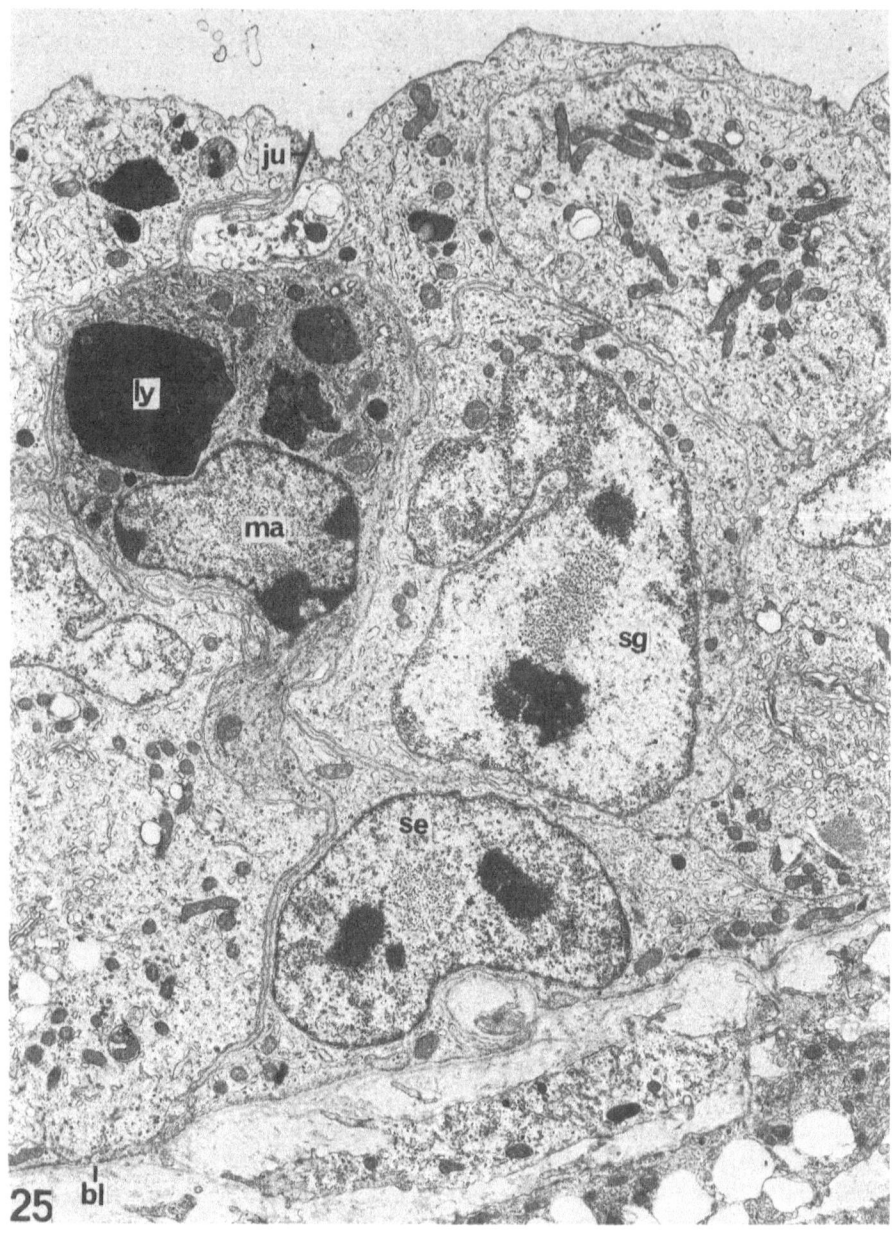

Fig. 25. Section of a seminiferous tubule at the end of August. Extended contact between macrophage and Sertoli cell. *bl*, basal lamina; *ju*, membrane junction between Sertoli cells; *ly*, lysosome; *ma*, macrophage; *se*, Sertoli cell; *sg*, spermatogonium. × 6700

Between the two extremes, the light and the dark Sertoli cells, there are numerous intermediate forms. Occasionally very differently appearing cells border on to one another (Fig. 24). If this is a case of different functional stages of one cell type, these are undergone independently of each other and of the neighboring germ cells.

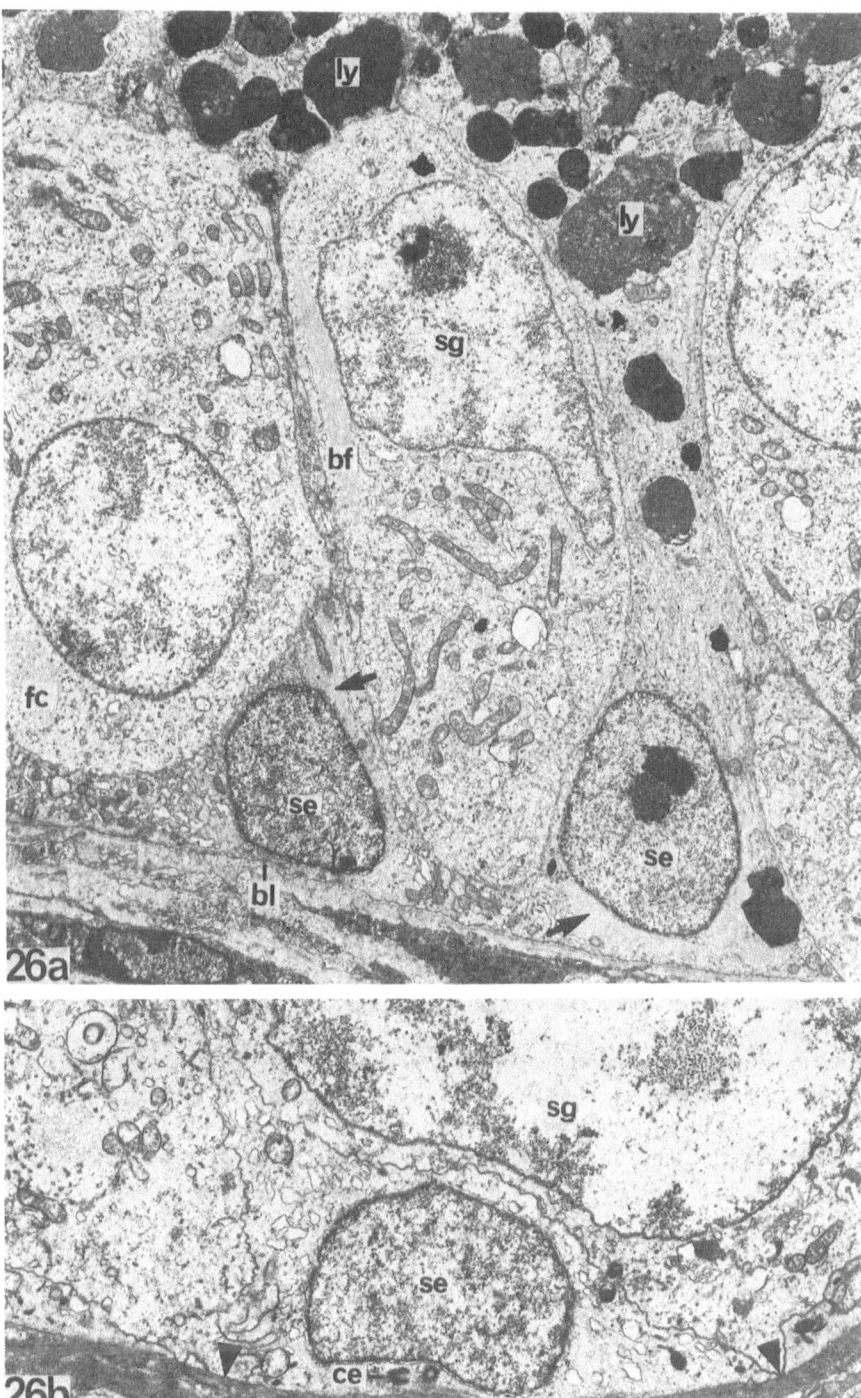

Fig. 26a, b. *a* Germinal epithelium from the end of the involution phase at the beginning of November. The nuclei of Sertoli cells are located near the basal cell membrane. *b* Sertoli cell from the resting phase at the middle of December. At the end of phagocytotic activity diplosomes are rather frequent. *bf*, bundles of thicker filaments; *bl*, basal lamina; *ce*, centriole; *fc*, filamentous-granular complex; *ly*, lysosome; *se*, Sertoli cell; *sg*, spermatogonium; *arrows*, thin filaments of Sertoli cells; *arrowheads*, contact of spermatogonium with the basal lamina. *a* × 5400, *b* × 7800

In addition to the cell components already described, there are two further striking characteristics which appear predominantly in the light Sertoli cells, but also in the dark ones. One of these is the centrioles (Fig. 26b), which appear as single entities or more often as diplosomes, and which are not to be found in the other phases of the annual cycle except occasionally in the resting phase in winter. Mitotic divisions in the Sertoli cells, however, have not been observed with certainty. On the contrary, there are sometimes two nuclei present in one cell. Moreover, some cells exhibit, either directly adjacent to the nucleus or supranuclearly in the vicinity of larger inclusions, a more or less extended area which is free from cell organelles and contains instead very slender microfilaments. These filaments are electron-optically unusually poor in contrast and one can only just recognize an approximately parallel orientation (Fig. 26a). It is noticeable also that in the involution phase glycogen never appears in the Sertoli cells of the swan, and fat droplets hardly ever.

8.3 Macrophages

Phase 4 of the annual cycle is particularly characterized by the massive invasion of free macrophages into the seminiferous tubules (Figs. 24, 27a). These cells are especially easy to identify because of certain characteristics when they appear in the very briefly present lumina of the tubules, but even so they can be recognized without any doubt on their way from the interstitial tissue across the germinal epithelium.

The free macrophage in the seminiferous tubule is very similar in its ultrastructure to the peritoneal macrophage in higher vertebrates. The more or less spherical cell exhibits on its surface finger-like or long thread-like pseudopodia (Figs. 24, 27a, 28a), which are made up of ectoplasm. These cell processes are mostly bent or extremely twisted; they can be branched and are frequently flattened back with their ends against the cell surface (Figs. 27a, 28a). Moreover, the surface sometimes exhibits deep indentations which give, depending on the cutting direction, the appearance of vacuoles in the peripheral cytoplasm (Figs. 24, 27a).

The nucleus of the macrophage is round to oval (Figs. 25, 27a, 28a) but can also be indented on one side (Figs. 25, 27b, 28b), and mostly contains a nucleolus. The karyoplasm shows a granulation which is somewhat irregularly distributed over the nuclear space, and a strikingly dense layer of chromatin along the nuclear membrane (Figs. 27a, b), which is more prominent than in the nuclei of the Sertoli cells and to which some larger clumps of chromatin may be attached (Figs. 25, 27a). Through this nuclear morphology it is possible to differentiate the macrophage clearly from the Sertoli cell, especially when the macrophage is found on its way across the germinal epithelium (Figs. 25, 27b, 30b).

The cytoplasmic matrix of the macrophages is often very dense, and mostly higher in contrast than that of the germ cells and that of nearly all Sertoli cells. The organelles are located in small cytoplasmic areas not occupied by phagocytosed material and generally do not advance into the pseudopodia (Figs. 27a, 28a). The few oval mitochondria are relatively small and contain in their likewise dense matrix parallel positioned cristae. With the increasing expansion of incorporated foreign bodies the Golgi apparatus becomes less and less distinct. The endoplasmic reticulum is present with and without attachment of ribosomes (Figs. 28a, 29a, b); the numerous free ribosomes are found mostly in the form of polysomes. Microtubules and microfilaments in grow-

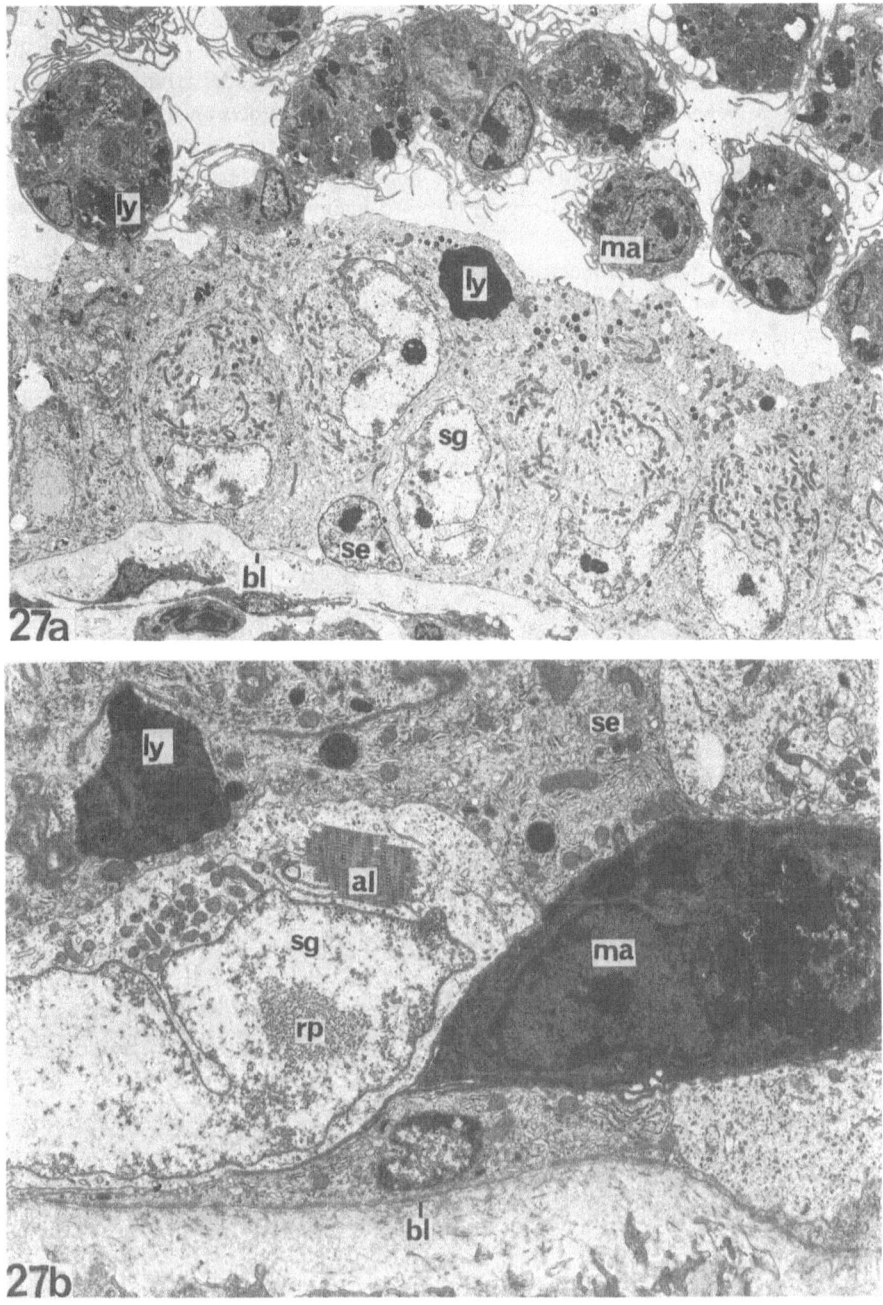

Fig. 27a, b. Sections from seminiferous tubules at the height of involution during invasion of macrophages at the end of August. Noteworthy are the broad areas of contact between the migrating macrophage and the Sertoli cells, and also the annulate lamellae in a spermatogonium. *al*, annulate lamellae; *bl*, basal lamina; *ly*, lysosome; *ma*, macrophage; *rp*, ribonucleoprotein complex; *se*, Sertoli cell; *sg*, spermatogonium. *a* × 2100, *b* × 6700

ing number and density can be observed predominantly in the peripheral areas of the cytoplasm (Figs. 27a, 29a, b); they have a diameter between 5 and 7 nm. Occasionally they are joined together to form bundles, and suggest a periodic arrangement of incomplete cross-links. In the lighter macrophages there sometimes appear round to oval structures, surrounded by a membrane and of differing sizes, which are presumably primary lysosomes. These multivesicular or homogeneous electron-dense bodies are frequently to be observed in close proximity to a somewhat more prominent Golgi apparatus.

Most striking about the macrophages is the considerable amount of incorporated and partly digested matter. This comes from degenerated germ cells, the nuclei of which are the most resistant structures during the breaking-down process (Fig. 28a). A macrophage frequently contains more than one phagocytosed germ cell (Figs. 28a, b). Even within the germinal epithelium degenerated germ cells as well as morphologically intact cells suffer phagocytosis (Figs. 28a, b).

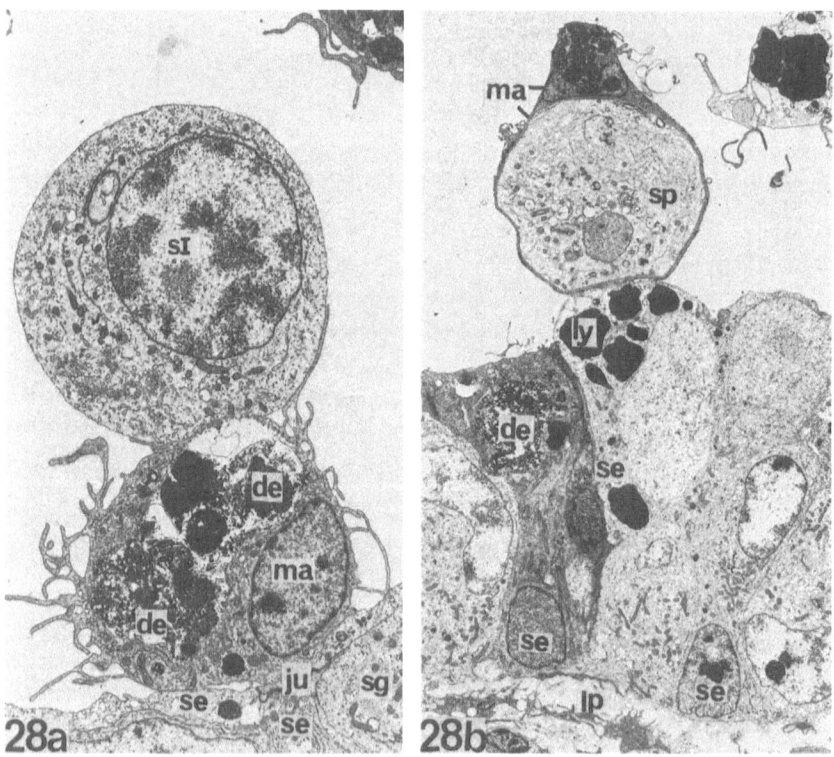

Fig. 28a, b. Incorporation of a primary spermatocyte *a* and a spermatid *b* into macrophages at the end of August. *de*, remnants of degenerated germ cells; *ju*, membrane junction between Sertoli cells; *lp*, lamina propria; *ly*, lysosome; *ma*, macrophage; *se*, Sertoli cell; *sg*, spermatogonium; *sI*, primary spermatocyte; *sp*, spermatid. *a* × 3850, *b* × 1650

Fig. 29a, b. Macrophages coming into contact with cells of the germinal epithelium at the end of August. *ju*, membrane junction between Sertoli cells; *ly*, lysosome; *ma*, macrophage; *ps*, pseudopodia of macrophage; *se*, Sertoli cell; *sg*, spermatogonium; *asterisk*, residual of germ cell; *arrows*, fine granular material attached to the cell membrane of the macrophage; *arrowheads*, area of fusion between macrophage and Sertoli cell. *a* × 51 200, *b* × 20 200

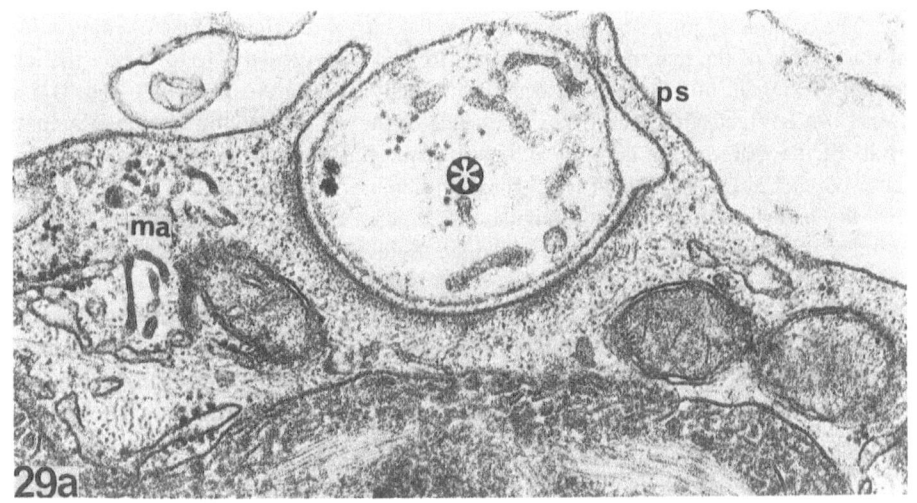

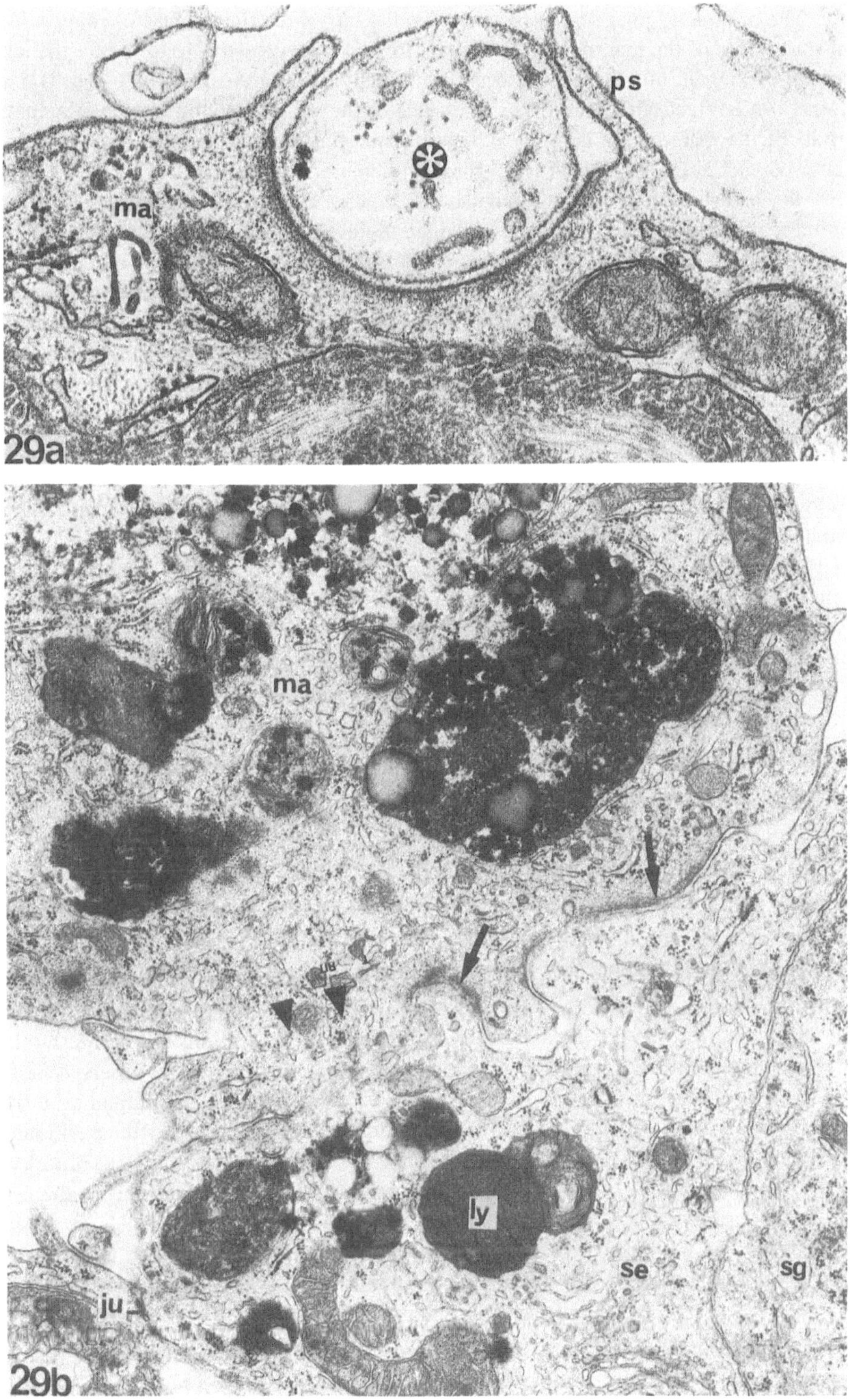

29a

29b

The process of phagocytosis can be divided into three steps. The close apposition of the surface of the macrophage to the cell to be phagocytosed is followed by the taking up of the cell into the macrophage, and finally by its digestion. The two cells first come into contact with each other by means of the exploratory outstretched pseudopodia of the macrophage (Fig. 27a). Cytoplasmic densities can frequently be found at these contact zones just underneath the cell membrane of both cells (Fig. 29a). With this the incorporation of the foreign body begins. The contact zone, in which the pseudopodia are retracted, becomes extended as does the zone of increased cytoplasmic density which is, however, restricted to a thin fringe underneath the cell membrane of the macrophage (Figs. 29a, b). Next pseudopodia protrude on both sides of the contact zone and extend along the circumference of the particle or the cell to be phagocytosed (Figs. 28a, 29a). Finally they meet and fuse with each other (Fig. 28b). The cell to be removed is thereby completely encompassed in a huge vacuole and the digestion can begin.

The mechanism which contributes to the more or less complete degradation of the incorporated materials cannot be morphologically determined in detail. One can only follow single steps of advancing cell degeneration inside the digestion vacuole. This entails the intensive absorption and activity of lysosomal enzymes, although the fusion of lysosome and phagosome is not observed very often. Also in the incorporated germ cell one mostly finds a changing number of different inclusions pointing to disintegration, such as myelin-like membrane complexes, multivesicular bodies, and vesicles of different size with homogeneous, electron-dense contents, which are probably primary lysosomes (Fig. 28b).

In advanced digestion the macrophage contains one or more large, extraordinarily heterogeneous inclusion bodies whose single cellular or lysosomal components are often no longer to be identified. Their content consists of areas of fine to very rough granular material, multivesicular bodies, filaments, the remains of membranes, and entities of different shapes and sizes which are similar to fat droplets (Figs. 25, 27a). All these components are surrounded by a membrane (Figs. 28b, 29b), which represents the limiting outline of the digestive vacuole but may also be the remains of the cell membrane of the degenerated germ cell.

The single steps of phagocytosis are especially easy to follow when cells which lie free in the lumen of the tubule or in the uppermost layer of the germinal epithelium undergo phagocytosis. These are the apparently favored locations where most cells undergo degeneration. Inside the germinal epithelium this process is observed more rarely, but occurs in principle in the same way. However, the pseudopodia of the macrophage cannot spread out so unhindered here, and are therefore often positioned in several layers one on top of the other (Figs. 25, 30a, b). The outhermost processes take over the function of incorporation, becoming very closely connected to the cell to undergo phagocytosis, which is indented and deformed by the advancing process. Even when the contact is so close that one can no longer make out the intercellular space, a disintegration of the cell membrane does not take place until the germ cell is completely enclosed by the macrophage.

In the contact zones between macrophages and Sertoli cells on the other hand, interruptions of the cell membranes do seem to appear, though only for a short distance. Such points of fusion can be seen in particular where a macrophage contacts the luminal free surface of a Sertoli cell (Figs. 28a, 29b). Whether an exchange of substances

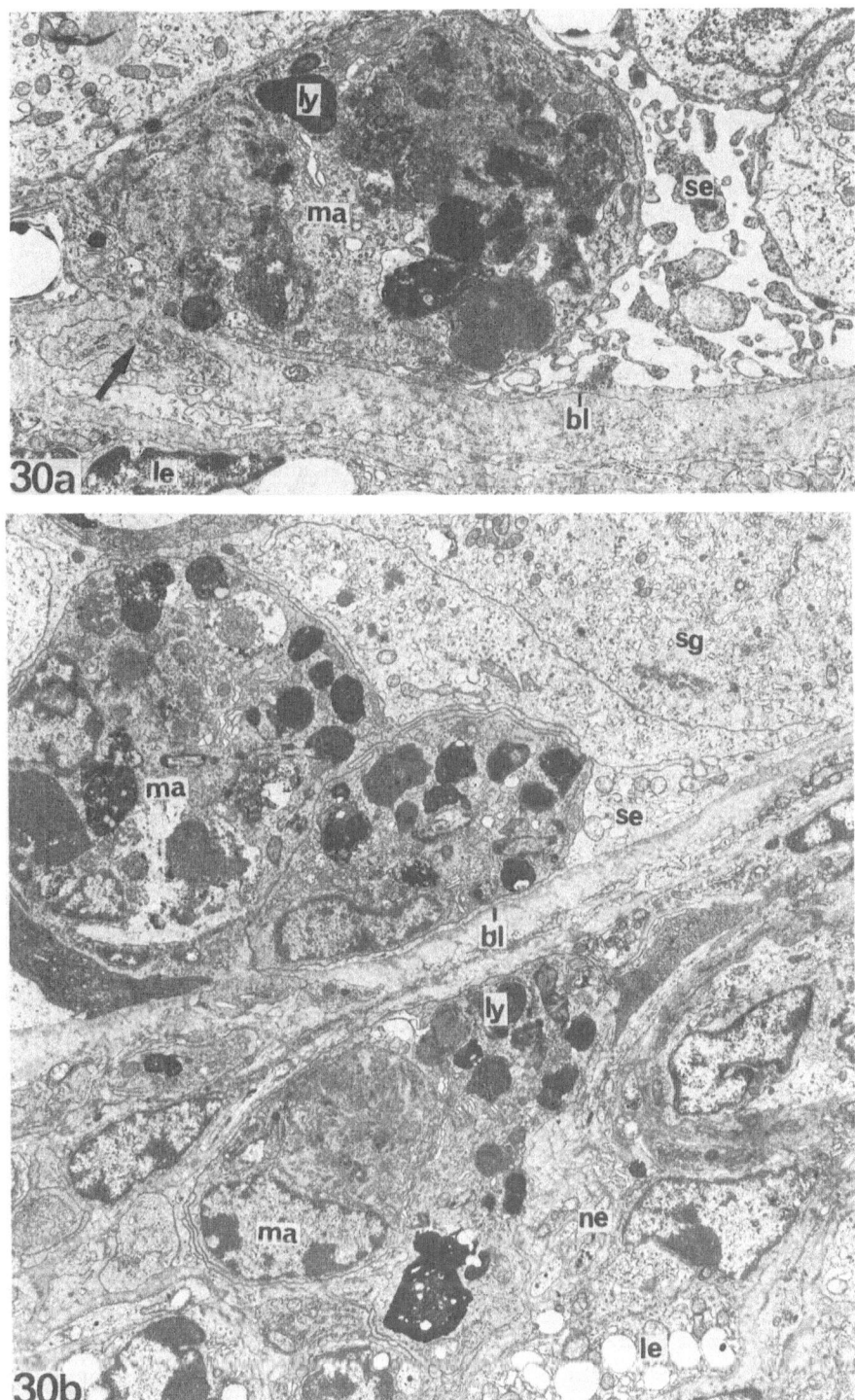

Fig. 30a, b. Macrophages on their way across the germinal epithelium, just passing the basal lamina (*arrow*), and in the interstitium *b*. *bl*, basal lamina; *le*, Leydig cell; *ly*, lysosome; *ma*, macrophage; *ne*, nerve fiber; *se*, Sertoli cell; *sg*, spermatogonium. *a* × 5400, *b* × 5600

takes place here is not clear; in both cell types similar inclusions are present because of a foregoing phagocytosis.

The majority of the macrophages leave the seminiferous tubules via the lumina, which briefly reappear in this phase, and through the rete testis. Some of them, however, return through the germinal epithelium into the interstitial tissue, since there one can find in addition to still inactive macrophages cells which have already taken part in phagocytosis (Fig. 30b). During their migration through the germinal epithelium, the macrophages do not reveal in which direction they are moving (Figs. 25, 27b, 30a). In their shape they fit into their surroundings and stretch out their pseudopodia towards the center of the tubule as well as towards the basal lamina (Fig. 25). They undergo a particularly strong deformation during their passage of the base of the tubule (Figs. 30a, b). As the spermatogonia seldom lie on the basal lamina with a broad base, in general two neighboring Sertoli cells, which have not developed any special cell contact in the basal area, are separated from another by the macrophage moving out or moving in (Fig. 30b). Consequently also inside the germinal epithelium one can frequently see large areas of contact between macrophage and Sertoli cell (Fig. 25). It appears as if the macrophage is pushing itself alongside the Sertoli cell.

8.4 Discussion

The degeneration of germ cells in the process of spermatogenesis is a physiological procedure widespread in the animal kingdom, and can regularly be observed not only in the adult testis (Roosen-Runge 1973) but also in the development of the fetal gonads (Black 1971). This is also true of the swan in the phase of full spermatogenic activity in spring (see p. 43). There is an increasing disintegration of germ cells under experimental or pathological changes of the testes, as has been described, for example, in higher vertebrates (Paufler and Foote 1969; Crabo et al. 1971; Marchand and Gomot 1976; Holstein 1978). However, a germ cell loss can also occur in consequence of naturally changed physiological conditions, as, for example, age-dependent alterations in the gonads (Holstein 1978). In seasonally breeding animals, the yearly involution of the testes exhibits an extreme physiological situation, in which especially large numbers of unusual germ cells at all stages of maturity degenerate or are disposed of (Miller 1948; Lofts et al. 1966b; Billard et al. 1972; Glover 1973; Marchand and Gomot 1973b; Chan and Lofts 1974).

The removal of degenerating or superfluous cells occurs as a rule either through the phagocytotic activity of the Sertoli cells observed by means of light and electron microscopy (Miller 1948; Roosen-Runge 1955; Billard et al. 1972; Schulze 1974; Marchand and Gomot 1976), or through the excurrent genital pathways from the rete testis to the vas deferens, whose epithelia are also capable of phagocytosis (Roussel et al. 1967; Tingari and Lake 1972a; Glover 1973; Burgos and Cavicchia 1975; Dym 1976; Cooper and Hamilton 1977). In the swan the involution phase ranges over several months, during which the seminiferous tubules are only seldom canalized, so that the transport of degenerated germ cells via the lumina generally does not come into question. Therefore the Sertoli cells take over this function, in that in comparison with the remaining seasons they develop an extremely increased phagocytotic activity.

In spite of the increased disintegration of degenerating germ cells the Sertoli cells do not show any fundamental changes in their branched structure, their nuclear morphology or in the ultrastructure of their cell organelles during the involution phase. However, it is conspicuous that the density of the nucleus and the cytoplasm in a number of cells increases to a degree hardly to be observed in the other phases of the annual cycle. Such differences in density have so far not received much attention except in developmental stages of the mammalian testis. Black and Christensen (1969) described in the fetal testis of the guinea-pig the different distribution of light and dark Sertoli cells, without giving it any special meaning. Wartenberg (1978) on the contrary maintained that the light and dark pre-Sertoli cells of the differentiating human testis represented two cell types with different origins and correspondingly different functions. These cells, however, differed from one another through the absence or presence of glycogen; they were thought to stimulate or inhibit the process of meiosis in the germ cells.

In comparison one can only speculate as to the meaning of the light and dark Sertoli cells in the swan. It is still unclear whether or not the dark Sertoli cells break down the phagocytosed material so completely that they finally become light colored again. In this case, the two extremes would represent only functional states of one cell type, and this is supported by the numerous intermediate forms which can be observed. On the other hand it is also feasible – and the Sertoli cell nuclei which occasionally lie very near the lumen speak for this idea – that the dark cells which are no longer present after the involution phase leave the germinal epithelium as spermatophages (Breucker 1975). Similar ideas were reflected by Phadke (1964), who derived the spermatophages in the human testis from the basal cells of the ductus epididymidis. His findings were based on light microscopic observations of paraffin sections. In the case of the swan, however, it would be necessary to postulate that the remaining Sertoli cells become replenished from a contingent of stem cells and that, contrary to the general assumption, they are in the adult still able to divide by mitosis, in order to secure a sufficient number of cells to take part in the next active phase of spermatogenesis. Really, from the involution phase until the resting phase a surprising number of diplosomes are present, predominantly in the light Sertoli cells, and there are also many cells with two nuclei. Clearly visible mitotic divisions have not so far been proved in the swan.

The additional participation of wandering cells such as macrophages in the disposal of degenerating germ cells has only been described in a few cases. These foreign cells, mostly described as spermatophages, predominantly remove mature or incompletely differentiated spermatozoa and their fragments. They appear, for example, in increasing numbers in the rete testis or epididymis in humans after vasectomy or ligature of the vas deferens (Phadke 1964), but are also present in the epididymis of healthy men (Holstein 1967, 1969). They can likewise be observed in the epididymis of the bull, rabbit, and monkey (Roussel et al. 1967) and also in the glandulae vesiculosae of the Japanese rhesus monkeys (Murakami et al. 1978). In the cock macrophages were found in the lumen and in the subepithelial connective tissue of the genital ducts following ligature of the vas deferens (Tingari and Lake 1972a). Under special physiological conditions, however, these cells can also be demonstrated in the seminiferous tubules and have been identified by means of electron microscopy as free invading macrophages. Thus Holstein (1978) could show in the human and the monkey that following captivity, ageing or immunological processes macrophages wander from the interstitial tissue across the epithelium of the seminiferous tubules and of the rete testis, in order to bring about phagocytosis of degenerating spermatozoa in the lumen.

In the monkey this happens mostly at the end of an active phase of spermatogenesis. A nearly comparable situation occurs in the involution phase of the seasonal cycle of the swan. The transient invasion of macrophages into the seminiferous tubules at the height of germ cell disintegration is a regular finding, which has not been described up to now in other seasonal breeders (Breucker 1978). There is only a notice mentioned by Wing and Lin (1977) that in the hamster macrophages in the interstitial tissue of the testis are exposed to annual changes in contact with Leydig cells, but without taking part in the breaking up of spermatozoa.

The macrophages in the swan can be followed on their way out of the interstitial blood-vessels through the lamina propria and the basal lamina of the seminiferous tubule, and across the germinal epithelium into the lumen of the tubule, and are therefore seen as foreign cells. In their ultrastructure they very much resemble the free macrophages which as mononuclear phagocytes descend from blood monocytes (van Furth et al. 1975), and which have been described in most different tissues, above all in mammals (Carr 1973; Nichols and Bainton 1975; van der Rhee et al. 1979). In the procedure of incorporation and the following lysosomal breaking down of picked up cells or of their fragments they also correspond to the generally accepted concept. Unlike the previously mentioned spermatophages of presumably the same origin (Holstein 1967, 1969; Tingari and Lake 1972a; Holstein 1978), the macrophages in the involution phase of the swan exclusively phagocytose immature germ cells, which are often found in a developmental state prior to termination of the meiotic divisions. In this way an elaborate differentiation of germ cells capable of fertilization, which are no longer needed during the present seasonal cycle, is prevented and the breakdown of the germinal epithelium before the resting phase is accelerated.

The macrophages in the swan begin their phagocytotic activity even during their passage through the germinal epithelium, so that they often appear laden with cell debris before they reach the lumen of the tubule. In spite of this they are still capable of further phagocytosis when they arrive. It is to be assumed that most of the macrophages leave the testis via the lumina of the seminiferous tubules and the excurrent ducts. However, some of them may appear to return through the germinal epithelium into the interstitium again. There they can be recognized because of signs of foregoing phagocytotic activity, such as remnants of incorporated primary spermatocytes in the case of the swan. In the human and the monkey, the macrophages with incorporated fragments of spermatozoa and spermatids usually pass through the rete epithelium (Holstein 1978). Moreover, after being expelled from the testis they can also cross the epithelium of the excurrent ducts, so that they can be found in the subepithelial connective tissue of the ductus epididymidis, together with their phagocytosed contents, as Tingari and Lake (1972a) have shown in the cock.

It may be supposed that the yearly occurrence of the macrophage invasion during the testicular regression is a finding which is not only confined to the swan. In other animal species with marked seasonal reproductive cycles, macrophages may presumably be present which have simply not been identified as such, but have been looked upon as phagocytosing Sertoli cells (Lofts et al. 1966b). It is true that the appearance of foreign cells in the seminiferous tubules during the regression phase has already occasionally been observed light microscopically, and their part in the clearing-up process has been suggested (polymorphonuclear leukocytes in the tree sparrow: Chan and Lofts 1974; macrophages and heterophil granulocytes in various species of blackbird: Payne 1969). A doubt-free identification is, however, only possible with a knowledge of the ultrastructure of the nucleus, for example. So also the electron micro-scopic findings of Glover (1973) suggest that the dark cells that he found in the epi-thelium of the epididymis in the rock hyrax shortly before the beginning of a new reproductive period could in fact have been macrophages taking part in the removal of old spermatozoa.

With regard to the swan, the question remains open by what means the macro-phages are stimulated suddenly to invade the seminiferous tubules in great numbers during the involution phase and to cooperate in the clearing-up process. The invasion sets in at a moment when the Sertoli cells are already overburdened with broken-down products. It is, moreover, striking that the macrophages very frequently take up contact with Sertoli cells on their way through the germinal epithelium as well as in the apical adluminal region of the latter. Therefore the Sertoli cells are thought to play an important role in the activation of the macrophages.

9 Summary

The changes in the annual course of spermatogenesis of a seasonal breeder, the mute swan, were examined by means of light and electron microscopy; this was taken as example of a barely domesticated avian species.

The maturation and differentiation of the male germ cells take place in the semini-ferous tubules as they do in all higher vertebrates. The complete process of spermato-genesis from spermatogonia to mature spermatozoa can only be observed during a few weeks in spring. In spite of the fact that synchronization of the single germ cell gener-ations is confined to small areas of the tubules, the development follows an ordering principle which can be subsumed under the term kinetics. By observing the nuclear morphology of the primary spermatocytes and the spermatids in semithin sections, the process of spermatogenesis in the swan can be divided into eight different cell associa-tions known as stages, whose chronological sequence represents a cycle of spermato-genesis. So far a statement about the stem cell renewal is not possible, because different cell classes of spermatogonia, i.e., reserve cells and cells entering maturation, cannot be differentiated morphologically.

The electron microscopic examination of the seminiferous tubules during the activity phase allows a detailed characterization of the single germ cell generations and of the Sertoli cells. Even in the area of ultrastructure there are no different types of spermatogonia. However, the single phases of the maturation divisions in the primary and secondary spermatocytes, and especially the stages of the extended first prophase, can be established exactly. During spermatid differentiation a series of characteristic structures and their changes are to be observed which are species-specific. Such a struc-ture are the annulate lamellae, which are regularly present already in the secondary spermatocytes and occasionally appear in spermatogonia; in the course of the sper-matid differentiation they make contact with different cell components: the endoplasmic reticulum, the nuclear membrane, the chromatoid body, and in particular a special tubular body. Further peculiarities of the spermatids are the rod-shaped structure in the area of the acrosome, the microtubular generations in the vicinity of the nucleus, a missing neck piece, and finally the persistent long-growing distal centriole. The Ser-toli cells represent a comparatively unstriking cell population during the active phase of spermatogenesis: they have an unlobed nucleus without signs of special activity and a normal content of cell organelles.

The seasonally related building up and breaking down of the germinal epithelium in the course of a year can be characterized by five especially marked phases. The annual cycle begins after the winter resting period with a multiplication phase of the spermatogonia. This passes into the activity phase, in which the peak of spermatogenesis is reached. This is followed by the involution phase, in which not only are maturation and differentiation inhibited, but also germ cells from all states of development degenerate and are to a large extent phagocytosed by the Sertoli cells. The climax of regression is marked by a special phase which is connected to a massive invasion of macrophages into the germinal epithelium. The end of the annual cycle occurs with the resting phase in winter.

The active invasion of macrophages from the interstitium into the germinal epithelium, and their participation in the breaking down of immature germ cells, is a finding recurrent in the swan but has escaped exact observation up to now in other animals with seasonal reproduction. In their ultrastructure these foreign cells as mononuclear phagocytes can be compared to the peritoneal macrophages. From electron micrographs one can follow their passage from the interstitium through the basal lamina of the seminiferous tubules and across the germinal epithelium. In the same way, their migration in the opposite direction back into the interstitium is possible, as subepithelial macrophages containing the remnants of phagocytosed germ cells demonstrate. The single steps of incorporation of germ cells, especially of spermatocytes and young spermatids, can be only observed within the scope of ultrastructure. During a short period of the regression phase the macrophages support the already maximally increased phagocytotic activity of the Sertoli cells. The two cell types often come into specially intense contact with one another. Their working together accelerates the clearing-up process during involution and possibly prevents the degeneration of Sertoli cells, which at the beginning of an annual cycle are again free from phagocytosed inclusions and in which, however, mitotic divisions cannot yet be observed.

References

Altorfer J, Fukuda T, Hedinger C (1974) Desmosomes in human seminiferous epithelium. Virchows Arch [Cell Pathol] 16:181–194

Aristoteles (1868) Thierkunde. Greek-German edition. Book III, Paragraph 5. Engelmann, Leipzig

Aschoff J (1955) Jahresperiodik der Fortpflanzung bei Warmblütern. Stud Gen 8:742–776

Baccetti B, Afzelius BA (1976) The biology of sperm cell. Karger, Basel

Ballowitz E (1888) Untersuchungen über die Struktur der Spermatozoen, zugleich ein Beitrag zur Lehre vom feineren Bau der kontraktilen Elemente. I. Die Spermatozoen der Vögel. Arch Mikr Anat 32:401–473

Bauer KM, Glutz von Blotzheim UN (1968) Handbuch der Vögel Mitteleuropas. Vol. II/1. Anseriformes. Akademische Verlagsgesellschaft, Frankfurt/M, pp 13–46

Baumgarten HG, Holstein AF (1968) Adrenerge Innervation im Hoden und Nebenhoden vom Schwan (Cygnus olor). Z Zellforsch 91:402–410

Baumgarten HG, Holstein AF (1974) Leydigzellinnervation und jahreszeitliche Schwankungen im Catecholamingehalt des Hodens von Vögeln. Verh Anat Ges 68:267–271

Beams HW, Kessel RG (1974) The problem of germ cell determinants. Int Rev Cytol 39:413–479

Benoit J (1936) Facteurs externes et internes de l'activite sexuelle. I. Stimulation par la lumière de l'activité sexuelle chez le canard et la cane domestique. Bull Biol Fr Belg 70:487–533

Benoit J (1950) Organes uro-génitaux. In: Grassé P (ed) Traité de Zoologie, vol XV Oiseaux. Masson, Paris, pp 350–355

Benvenuti C (1970) Risultati della ricostruzione totale dei tubuli seminiferi contorti di canarino (Serinus canarius) adulto in attività ed in riposo sessuale. Ann Fac Med Vet (Pisa) 23:288–306

Bernhard W (1968) Une méthode de coloration régressive à l'usage de la microscopie électronique. C R Acad Sci [D] (Paris) 267:2170–2173

Berthold P, Gwinner E, Klein H (1972) Circannuale Periodik bei Grasmücken. II. Periodik der Gonadengröße bei Sylvia atricapilla und S. borin unter verschiedenen konstanten Bedingungen. J Ornithol 113:407–417

Bigliardi E, Vegni Talluri M (1976) Ultrastructural details of Sertoli cell junctional complexes in vivo and their modifications in tissue culture. Cell Tissue Res 172:29–38

Billard R, Jalabert B, Breton B (1972) Les cellules de Sertoli des poissons téléostéens. I. Etude ultrastructurale. Ann Biol Anim Biochim Biophys 12:19–32

Bissonnette TH (1930) Studies on the sexual cycle in birds. I. Sexual maturity, its modification and possible control in the European starling (Sturnus vulgaris). Am J Anat 45:289–305

Bissonnette TH, Chapnick MH (1930) Studies on the sexual cycle in birds. II. The normal progressive changes in the testis from November to May in the European starling (Sturnus vulgaris), an introduced, non-migratory bird. Am J Anat 45:307–343

Black VH (1971) Gonocytes in fetal guinea pig testes: phagocytosis of degenerating gonocytes by Sertoli cells. Am J Anat 131:415–426

Black VH, Christensen AK (1969) Differentiation of interstitial cells and Sertoli cells in fetal guinea pig testes. Am J Anat 124:211–237

Blanchard BD (1941) The white-crowned sparrows (Zonotrichia leucophrys) of the pacific seaboard: environment and annual cycles. Univ Calif Publ Zool 46:1–178

Blanchard BD, Erickson MM (1949) The cycle in the gambel sparrow. Univ Calif Publ Zool 47:255–318

Breucker H (1975) Saisonbedingte Involution der Spermatogenese beim Schwan. Verh Anat Ges 69:733–737

Breucker H (1978) Macrophages, a normal component in seasonally involuting testes of the swan, Cygnus olor. Cell Tissue Res 193:463–471

Breucker H, Horstmann E (1968) Die Spermatozoen der Zecke Ornithodorus moubata (Murr). Z Zellforsch 88:1–22

Breucker H, Horstmann E (1972) Die Spermatogenese der Zecke Ornithodorus moubata (Murr). Z. Zellforsch 123:18–46

Brökelmann J (1963) Fine structure of germ cells and Sertoli cells during the cycle of the seminiferous epithelium in the rat. Z Zellforsch 59:820–850

Brown NL, Baylé JD, Scanes CG, Follet BK (1975) Chicken gonadotrophins: their effects on the testes of immature and hypohysectomized Japanese quail. Cell Tissue Res 156:499–520

Brunn A von (1884) Beiträge zur Kenntnis der Samenkörper und ihrer Entwicklung bei Säugetieren und Vögeln. Arch Mikr Anat 23:108–132

Bullough WS (1942) The reproductive cycles of the British and continental races of the starling (*Sturnus vulgaris* L.). Philos Trans R Soc Lond [Biol] 231:165–246

Burger JW (1948) The relation of external temperature to spermatogenesis in the male starling. J Exp Zool 109:259–266

Burgos MH, Cavicchia JC (1975) Phagocytic activity in the epithelium of the rete testis. Proc 10th Int Congr Anat, Tokyo. Science Council of Japan, p 444

Burgos MH, Fawcett DW (1955) Studies on the fine structure of the mammalian testis. I. Differentiation of the spermatids in the cat (*Felis domestica*). J Biophys Biochem Cytol 1:287–300

Burgos MH, Fawcett DW (1956) An electron microscope study of spermatid differentiation in the toad, *Bufo arenarum* Hensel. J Biophys Biochem Cytol 2:223–240

Bustos-Obregón E, Holstein AF (1973) On structural patterns of the lamina propria of human seminiferous tubules. Z Zellforsch 141:413–425

Carr I (1973) The macrophage. A review of ultrastructure and functions. Academic Press, New York London

Carr I, Clegg EJ, Meek GA (1968) Sertoli cells as phagocytes: an electron microscopic study. J Anat 102:501–509

Cavazos LF, Melampy RM (1954) A comparative study of periodic acid-reactive carbohydrates in vertebrate testes. Am J Anat 95:467–495

Chan KMB, Lofts B (1974) The testicular cycle and androgen biosynthesis in the tree sparrow (*Passer montanus saturatus*). J Zool (Lond) 172:47–66

Christensen AK (1965) Microtubules in Sertoli cells of the guinea pig testis. Anat Rec 151:335

Christensen AK, Fawcett DW (1966) The fine structure of testicular interstitial cells in mice. Am J Anat 118:551–572

Christensen AK, Mason NR (1965) Comparative ability of seminiferous tubules and interstitial tissue of rat testes to synthesize androgens from Progesterone-4-^{14}C in vitro. Endocrinology 76:646–656

Clark AW (1967) Some aspects of spermiogenesis in a lizard. Am J Anat 121:369–400

Clegg EJ, MacMillan EW (1965) The uptake of vital dyes and particulate matter by the Sertoli cells of the rat testis. J Anat 99:219–229

Clermont Y (1958) Structure de l'épithélium séminal et mode de renouvellement des spermatogonies chez le canard. Arch Anat Microsc Morphol Exp 47:47–66

Clermont Y (1963) The cycle of the seminiferous epithelium in man. Am J Anat 112:35–51

Clermont Y (1972) Kinetics of spermatogenesis in mammals: seminiferous epithelium cycle and spermatogonial renewal. Physiol Rev 52:198–236

Coleman JR, Moses MJ (1964) DNA and the fine structure of synaptic chromosomes in the domestic rooster (*Gallus domesticus*). J Cell Biol 23:63–78

Comings DE, Okada TA (1972) The chromatoid body in mouse spermatogenesis: evidence that it may be formed by the extrusion of nucleolar components. J Ultrastruct Res 39:15–23

Connell CJ (1978) A freeze-fracture and lanthanum tracer study of the complex junction between Sertoli cells of the canine testis. J Cell Biol 76:57–75

Cooksey EJ, Rothwell B (1973) The ultrastructure of the Sertoli cell and its differentiation in the domestic fowl (*Gallus domesticus*). J Anat 114:329–345

Cooper TG, Hamilton DW (1977) Phagocytosis of spermatozoa in the terminal region and gland of the vas deferens of the rat. Am J Anat 150:247–268

Cowles RB, Nordstrom A (1946) A possible avian analogue to the scrotum. Science 104:586–587

Crabo B, Gustafsson B, Nicander L, Rao AR (1971) Subnormal testicular function in a bull concealed by phagocytosis of abnormal spermatozoa in the efferent ductules. J Reprod Fertil 26:393–396

Dan J, Hagiwara Y (1967) Studies on the acrosome. IX. Course of acrosome reaction in the starfish. J Ultrastruct Res 18:562–579

Disselhorst R (1908) Gewichts- und Volumenszunahme der männlichen Keimdrüsen bei Vögeln und Säugern in der Paarungszeit; Unabhängigkeit des Wachstums. Anat Anz 32:113–117

Dubois R, Cuminge D (1979) Sur le déterminisme de l'asymétrie primaire de répartition des cellules germinales dans les ébauches gonadiques chez l'embryon de poulet. C R Acad Sci [D] (Paris) 288:895–898

Dufaure JP (1971) L'ultrastructure du testicule de lézard vivipare (Reptile, Lacertilien). II. Les cellules de Sertoli. Étude du glycogène. Z Zellforsch 115:565–578

Duncker HR (1971) The lung air sac system of birds. Ergeb Anat Entwicklungs-Gesch 45:1–171

Duncker HR (1979) Coelomic cavities. In: King AS, McLelland J (eds) Form and function in birds, vol I. Academic Press, New York London, pp 39–67

Dustin P (1978) Nuclear and cytoplasmic shaping in spermatogenesis. In: Dustin P Microtubules. Springer, Berlin Heidelberg New York

Dym M (1976) The mammalian rete testis: a morphological examination. Anat Rec 186:493–524

Dym M, Cavicchia JC (1977) Further observations on the blood-testis barrier in monkeys. Biol Reprod 17:390–403

Dym M, Fawcett DW (1970) The blood-testis barrier in the rat and the physiological compartmentation of the seminiferous epithelium. Biol Reprod 3:308–326

Eddy EM (1974) Fine structural observations on the form and distribution of nuage in germ cells of the rat. Anat Rec 178:731–758

Enders AC (1963) Fine structural studies of implantation in the armadillo. In: Enders AC (ed) Delayed implantation. Rice University Press, Chicago, pp 281–292

Engels WL, Jenner CE (1956) The effect of temperature on testicular recrudescence in juncos at different photoperiods. Biol Bull 110:129–137

Farner DS, King JR (1971) Preface. In: Farner DS, King JR, Parkes KC (eds) Avian biology, vol I. Academic Press, New York London

Farner DS, Lewis RA (1971) Photoperiodism and reproductive cycles in birds. In: Giese AC (ed) Photophysiology, vol VI. Academic Press, New York London, pp 325–370

Farner DS, Lewis RA (1973) Field and experimental studies of the annual cycles of white-crowned sparrows. J Reprod Fertil [Suppl] 19:35–50

Farner DS, Wilson AC (1957) A quantitative examination of testicular growth in the white-crowned sparrow. Biol Bull 113:254–267

Fawcett DW (1956) The fine structure of chromosomes in the meiotic prophase of vertebrate spermatocytes. J Biophys Biochem Cytol 2:403–406

Fawcett DW (1965) The anatomy of the mammalian spermatozoon with particular reference to the guinea pig. Z Zellforsch 67:279–296

Fawcett DW (1972) Observations on cell differentiation and organelle continuity in spermatogenesis. In: Beatty RA, Gluecksohn-Waelsch S (eds) Proc int symp. The genetics of the spermatozoon. University Press, Edinburgh New York, pp 37–68

Fawcett DW, Phillips DM (1969) The fine structure and development of the neck region of the mammalian spermatozoon. Anat Rec 165:153–184

Fawcett DW, Eddy EM, Phillips DM (1970) Observations on the fine structure and relationships of the chromatoid body in mammalian spermatogenesis. Biol Reprod 2:129–153

Fawcett DW, Anderson WA, Phillips DM (1971) Morphogenetic factors influencing the shape of the sperm head. Dev Biol 26:220–251

Flickinger C, Fawcett DW (1967) The junctional specializations of Sertoli cells in the seminiferous epithelium. Anat Rec 158:207–222

Folliot R, Maillet PL (1965) Sur un aspect fonctionnel transitoire du réticulum endoplasmique au cours de la spermatogenèse de Dysdercus fasciatus Sign. (Hemiptera, Pyrrhocoridae). C R Soc Biol (Paris) 159:2483–2484

Fouquet JP (1974) La spermiation et la formation des corps résiduels chez le hamster: rôle des cellules de Sertoli. J Microsc (Paris) 19:161–168

Furieri P (1962) Prime osservazioni al microscopio elettronico sull'ultrastruttura degli spermatozoi di Fringilla coelebs L. Boll Soc Ital Biol Sper 38:29–32

Furth R van, Langevoort HL, Schaberg A (1975) Mononuclear phagocytes in human pathology – proposal for an approach to improved classification. In: Furth R van (ed) Mononuclear phagocytes in immunity, infection, and pathology. Blackwell, Oxford London Edinburgh Melbourne, pp 1–15

Garnier DH, Tixier-Vidal A, Gourdji D, Picart R (1973) Ultrastructure des cellules de Leydig et des cellules de Sertoli au cours du cycle testiculaire du canard Pékin. Corrélation avec les données biochimiques et cytoenzymologiques. Z Zellforsch 144:369–394

Glover TD (1973) The place of the seasonal breeder in research on male reproduction. In: Raspé G, Bernhard S (eds) Advances in the biosciences, vol 10, Schering workshop on contraception: the masculine gender. Pergamon Press, Oxford Edinburgh New York Toronto Sydney, Vieweg, Braunschweig–Wiesbaden, pp 235–246

Grigg GW, Hodge AJ (1949) Electron microscopic studies of spermatozoa. I. The morphology of the spermatozoon of the common domestic fowl (Gallus domesticus). Aust J Sci Res [B] 2: 271:286

Grzimek B, Meise W, Niethammer G, Steinbacher J, Thenius E (eds) (1968) Grzimeks Tierleben, vol 7/1. Kindler, Zürich

Gunawardana VK (1977) Stages of spermatids in the domestic fowl: a light microscope study using Araldite sections. J Anat 123:351–360

Gunawardana VK, Scott MGAD (1977) Ultrastructural studies on the differentiation of spermatids in the domestic fowl. J Anat 124:741–755

Guyer MF (1909a) The spermatogenesis of the domestic guinea (Numida meleagris dom.). Anat Anz 34:502–513

Guyer MF (1909b) The spermatogenesis of the domestic chicken (Gallus gallus dom.). Anat Anz 34:573–580

Gwinner E (1973) Circannual rhythms in birds: their interaction with circadian rhythms and environmental photoperiod. J Reprod Fertil (Suppl) 19:51–65

Haase E (1973) Zur Kontrolle von Fortpflanzungszyklen bei Vögeln: Untersuchungen an Bergfinken. J Comp Physiol 84:375–431

Hamner WM (1966) Photoperiodic control of the annual testicular cycle in the house finch, Carpodacus mexicanus. Gen Comp Endocrinol 7:224–233

Herin RA, Booth NH, Johnson RM (1960) Thermoregulatory effects of abdominal air sacs on spermatogenesis in domestic fowl. Am J Physiol 198:1343–1345

Hiatt RW, Fisher HJ (1947) The reproductive cycle of ring-necked pheasants in Montana. Auk 64: 528–548

Hilprecht A (1970) Höckerschwan, Singschwan, Zwergschwan. Neue Brehm-Bücherei 177, 2nd edn. Ziemsen, Wittenberg Lutherstadt

Holstein AF (1967) Spermiophagen im Nebenhoden des Menschen. Naturwissenschaften 54:98–99

Holstein AF (1969) Morphologische Studien am Nebenhoden des Menschen. Zwanglose Abhandl Gebiet Norm Pathol Anat (Stuttgart) vol 20, pp 1–91

Holstein AF (1978) Spermatophagy in the seminiferous tubules and excurrent ducts of the testis in rhesus monkey and man. Andrologia 10:331–352

Holstein AF, Wartenberg H (1976) Zur Entwicklung der Schwanzstrukturen bei Spermatiden des Menschen. Verh Anat Ges 70:875–880

Horstmann E (1961) Elektronenmikroskopische Untersuchungen zur Spermiohistogenese beim Menschen. Z Zellforsch 54:68–89

Horstmann E (1970) Structures of caryoplasm during differentiation of spermatids. In: Holstein AF, Horstmann E (eds) Morphological aspects of andrology. Grosse, Berlin, pp 24–28

Humphreys PN (1972) Brief observations on the semen and spermatozoa of certain passerine and non-passerine birds. J Reprod Fertil 29:327–336

Humphreys PN (1975a) The differentiation of the acrosome in the spermatid of the budgerigar (Melopsittacus undulatus). Cell Tissue Res 156:411–416

Humphreys PN (1975b) Ultrastructure of the budgerigar testis during a photoperiodically induced cycle. Cell Tissue Res 159:541–550

Immelmann K (1971) Ecological aspects of periodic reproduction. In: Farner DS, King JR, Parkes KC (eds) Avian biology, vol I. Academic Press, New York London, pp 341–389

Ito S, Winchester RJ (1963) The fine structure of the gastric mucosa in the bat. J Cell Biol 16: 541–577

Johnson OW (1961) Reproductive cycle of the mallard duck. Condor 63:351–364

Johnson OW (1966) Quantitative features of spermatogenesis in the mallard (Anas platyrhynchos). Auk 83:233–239

Johnston DW (1956) The annual reproductive cycle of the California gull. I. Criteria of age and the testis cycle. Condor 58:134–162

Kalt MR (1973) Ultrastructural observations on the germ line of *Xenopus laevis*. Z Zellforsch 138: 41–62

Kaya M, Harrison RG (1976) The ultrastructural relationships between Sertoli cells and spermatogenic cells in the rat. J Anat 121:279–290

Keast JA, Marshall AJ (1954) The influence of drought and rainfall on reproduction in Australian desert birds. Proc Zool Soc (Lond) 124:493–499

Kessel RG (1967) An electron microscope study of spermiogenesis in the grasshopper with particular reference to the development of microtubular systems during differentiation. J Ultrastruct Res 18:677–694

Kessel RG (1968) Annulate lamellae. J Ultrastruct Res Suppl 10:1–82

Kessel RG (1970) Spermiogenesis in the dragonfly with special reference to a consideration of the mechanisms involved in the development of cellular asymmetry. In: Baccetti B (ed) Comparative spermatology. Academic Press, New York London, pp 531–552

King AS (1975) Aves urogenital system. The male genital organs. In: Getty R (ed) Sisson and Grossman's the anatomy of the domestic animals, vol 2. Saunders, Philadelphia London Toronto, pp 1927–1935

Kirschbaum A, Ringoen AR (1935/1936) Seasonal sexual activity and its experimental modification in the male sparrow, *Passer domesticus* Linnaeus. Anat Rec 64 (Suppl):453–473

Koelliker RA von (1841) Beiträge zur Kenntnis der Geschlechtsverhältnisse und der Samenflüssigkeit wirbelloser Thiere. Logier, Berlin

Kretser DM de (1969) Ultrastructural features of human spermiogenesis. Z Zellforsch 98:477–505

Lacy D (1960) Light and electron microscopy and its use in the study of factors influencing spermatogenesis in the rat. J R Mircrosc Soc 79:209–225

Lacy D (1962) Certain aspects of testis structure and function. Br Med Bull 18:205–208

Lake PE (1957) The male reproductive tract of the fowl. J Anat 91:116–129

Lake PE, Smith W, Young D (1968) The ultrastructure of the ejaculated fowl spermatozoon. Q J Exp Physiol 53:356–366

Lanzavecchia G, Lamia Donin C (1972) Morphogenetic effects of microtubules. II. Spermiogenesis in *Lumbricus terrestris*. J Submicrosc Cytol 4:247–260

Leblond CP, Clermont Y (1952a) Definition of stages of the cycle of the seminiferous epithelium in the rat. Ann NY Acad Sci 55:548–573

Leblond CP, Clermont Y (1952b) Spermiogenesis of rat, mouse, hamster, and guinea pig as revealed by the "periodic-acid-fuchsin sulferous acid" technique. Am J Anat 90:167–215

Leuckart R (1853) Zeugung. In: Wagner's Handwörterbuch der Physiologie, vol IV (cited by Disselhorst und Schweigger-Seidel)

Lloyd HG, Englund J (1973) The reproductive cycle of the red fox in Europe. J Reprod Fertil [Suppl] 19:119–130

Lofts B (1962) Cyclical changes in the interstitial and spermatogenetic tissue of migratory waders wintering in Africa. Proc Zool Soc (Lond) 138:405–413

Lofts B (1964) Evidence of an autonomous reproductive rhythm in an equatorial bird (*Quelea quelea*). Nature 201:523–524

Lofts B (1970) Cytology of the gonads and feed-back mechanism with respect to photosexual relationships in male birds. Colloq Int Cent Nat Rech Sci 172:307–324

Lofts B (1972) The Sertoli cell. Gen Comp Endocrinol [Suppl] 3:636–648

Lofts B, Choy LYL (1971) Steroid synthesis by the seminiferous tubules of the snake *Naja naja*. Gen Comp Endocrinol 17:588–591

Lofts B, Coombs CJF (1965) Photoperiodism and the testicular refractory period in the mallard. J Zool (Lond) 146:44–54

Lofts B, Murton RK (1966) The role of weather, food and biological factors in timing the sexual cycle of wood-pigeons. Br Birds 59:261–280

Lofts B, Murton RK (1968) Photoperiodic and physiological adaptations regulating avian breeding cycles and their ecological significance. J Zool (Lond) 155:327–394

Lofts B, Murton RK (1973) Reproduction in birds. In: Farner DS, King JR, Parkes KC (eds) Avian biology, vol III. Academic Press, New York London, pp 1–107

Lofts B, Murton RK, Westwood NJ (1966a) Gonad cycles and the evolution of breeding seasons in British *Columbidae*. J Zool (Lond) 150:249–272

Lofts B, Phillips JG, Tam WH (1966b) Seasonal changes in the testis of the cobra, *Naja naja* (Linn.). Gen Comp Endocrinol 6:466–475

Marchand CR (1973) Ultrastructure des cellules de Leydig et des cellules de Sertoli du testicule du canard de Barbarie (*Cairina moschata* L.) en activité sexuelle. C R Soc Biol (Paris) 167:933–937

Marchand CR (1977) Étude ultrastructurale de la spermatogenèse du canard de Barbarie (*Cairina moschata* L., oiseau anatidé). Biol Cell 29:193–202

Marchand CR, Gomot L (1973a) Étude histologique et cytologique des testicules et des voies génitales du canard de Barbarie (*Cairina moschata* L.) en activité sexuelle. Journ Recherches Avicoles Cunicoles (Paris) 1973, 127–134

Marchand CR, Gomot L (1973b) Le cycle testiculaire du canard de Barbarie (*Cairina moschata* L.). Bull Assoc Anat (Nancy) 57:367–374

Marchand CR, Gomot L (1976) Spermatogenèse abortive chez le canard hybride (du croisement *Anas platyrhynchos* mâle x *Cairina moschata* femelle). Étude ultrastructurale. Bull Assoc Anat (Nancy) 60:613–622

Marchand CR, Gomot L, Reviers M de (1977) Étude par autoradiographie et marquage à la thymidine tritée de la durée de la spermatogenèse du canard de Barbarie (*Cairina moschata* L.). C R Soc Biol (Paris) 171:927

Maretta M (1975a) The ultrastructure of the spermatozoon of the drake. I. Head. Acta Vet Acad Sci Hung 25:47–52

Maretta M (1975b) The ultrastructure of the spermatozoon of the drake. II. Tail. Acta Vet Acad Sci Hung 25:53–60

Maretta M (1977) The behaviour of centrioles and the formation of the flagellum in rooster and drake spermatids. Cell Tissue Res 176:265–273

Marshall AJ (1949a) On the function of the interstitium of the testis. The sexual cycle of a wild bird, *Fulmaris glacialis* (L.). Q J Microsc Sci 90:265–280

Marshall AJ (1949b) Weather factors and spermatogenesis in birds. Proc Zool Soc (Lond) 119:711–716

Marshall AJ (1961a) Reproduction. In: Marshall AJ (ed) Biology and comparative physiology of birds, vol II. Academic Press, New York London, pp 169–213

Marshall AJ (1961b) Breeding seasons and migration. In: Marshall AJ (ed) Biology and comparative physiology of birds, vol II. Academic Press, New York London, pp 307–339

Marshall AJ (1970) Environmental factors other than light involved in the control of sexual cycles in birds and mammals. Colloq Int Cent Nat Rech Sci 172:53–64

Marshall AJ, Coombs CJF (1957) The interaction of environmental, internal and behavioural factors in the rook, *Corvus F. frugilegus* Linnaeus. Proc Zool Soc (Lond) 128:545–589

Marshall AJ, Roberts JD (1959) The breeding biology of equatorial vertebrates: reproduction of cormorants (*Phalacrocoracidae*) at latitude 0° 20′N. Proc Zool Soc (Lond) 132:617–625

Marshall AJ, Serventy DL (1957) On the postnuptial rehabilitation of the avian testis tunic. Emu 57:59–63

Marshall AJ, Serventy DL (1958) The internal rhythm of reproduction in xerophilous birds under conditions of illumination and darkness. J Exp Biol 35:666–670

Marshall AJ, Serventy DL (1959) Experimental demonstration of an internal rhythm of reproduction in a trans-equatorial migrant (the short-tailed shearwater *Puffinus tenuirostris*). Nature 184:1704–1705

Mattei C, Mattei X, Manfredi JL (1972) Electron microscope study of the spermiogenesis of *Streptopelia roseogrisea*. J Submicrosc Cytol 4:57–73

McIntosh JR, Porter KR (1967) Microtubules in the spermatids of the domestic fowl. J Cell Biol 35:153–173

McLelland J, King AS (1970) The gross anatomy of the peritoneal coelomic cavities of *Gallus domesticus*. Anat Anz 127:480–490

Menaker M (1971) Rhythms, reproduction and photoreception. Biol Reprod 4:295–308

Miller AH (1955) Breeding cycles in a constant equatorial environment in Columbia, South America. Experientia [Suppl] 3:495–503

Miller AH (1959) Reproductive cycles in an equatorial sparrow. Proc Natl Acad Sci USA 45:1095–1100

Miller MR (1948) The seasonal histological changes occurring in the ovary, corpus luteum, and testis of the viviparous lizard, *Xantusia vigilis*. Univ Calif Publ Zool 47:197–216

Miller RA (1938) Spermatogenesis in a sex-reversed female and in normal males of the domestic fowl, *Gallus domesticus*. Anat Rec 70:155–189

Mori JG, George JC (1978) Seasonal histological changes in the gonads, thyroid and adrenal of the Canada goose (*Branta canadensis interior*). Acta Anat (Basel) 101:304–324

Moses MJ (1956) Chromosomal structures in crayfish spermatocytes. J Biophys Biochem Cytol 2:215–218

Moses MJ (1958) The relation between the axial complex of meiotic prophase chromosomes and chromosome pairing in a salamander (*Plethodon cinereus*). J Biophys Biochem Cytol 4:633–638

Murakami M, Sugita A, Shimada T, Yoshimura T (1978) Scanning electron microscope observation of the seminal vesicle in the Japanese monkey with special reference to intraluminal spermiophagy by macrophages. Arch Histol Jpn 41:275–283

Murton RK, Kear J (1973) The nature and evolution of the photoperiodic control of reproduction in wildfowl of the family *Anatidae*. J Reprod Fertil [Suppl] 19:67–84

Murton RK, Westwood NJ (1977) Avian breeding cycles. Clarendon, Oxford

Murton RK, Lofts B, Westwood NJ (1970) The circadian basis of photoperiodically controlled spermatogenesis in the greenfinch *Chloris chloris*. J Zool (Lond) 161:125–136

Nagano T (1959) Spermatogenesis of the domestic fowl studied with the electron microscope. Arch Histol Jpn 16:311–345

Nagano T (1961) The structure of cytoplasmic bridges in dividing spermatocytes of the rooster. Anat Rec 141:73–79

Nagano T (1962) Observations on the fine structure of the developing spermatid in the domestic chicken. J Cell Biol 14:193–205

Nebel BR, Coulon EM (1962) The fine structure of chromosomes in pigeon spermatocytes. Chromosoma 13:272–291

Nicander L (1967) An electron microscopical study of cell contacts in the seminiferous tubules of some mammals. Z Zellforsch 83:375–397

Nicander L (1970) Comparative studies on the fine structure of vertebrate spermatozoa. In: Baccetti B (ed) Comparative spermatology. Academic Press, New York London, pp 47–55

Nichols BA, Bainton DF (1975) Ultrastructure and cytochemistry of mononuclear phagocytes. In: Furth R van (ed) Mononuclear phagocytes in immunity, infection, and pathology. Blackwell, Oxford London Edinburgh Melbourne, pp 17–55

Okamura F, Nishiyama H (1976) The early development of the tail and the transformation of the shape of the nucleus of the spermatid of the domestic fowl, *Gallus gallus*. Cell Tissue Res 169:345–359

Paufler SK, Foote RH (1969) Spermatogenesis in the rabbit following ligation of the epididymidis at different levels. Anat Rec 164:339–348

Payne RB (1969) Breeding seasons and reproductive physiology of tricolored blackbirds and red-winged blackbirds. Univ Calif Publ Zool 90:1–125

Phadke AM (1964) Fate of spermatozoa in cases of obstructive azoospermia and after ligation of vas deferens in man. J Reprod Fertil 7:1–12

Phillips DM (1970) Development of spermatozoa in the wooly opossum with special reference to the shaping of the sperm head. J Ultrastruct Res 33:369–380

Phillips DM (1974) Nuclear shaping in the absence of microtubules in scorpion spermatids. J Cell Biol 62:911–917

Picheral B (1972) Les éléments cytoplasmiques au cours de la spermiogenèse du triton *Pleurodeles waltlii* Michah. I. La genèse de l'acrosome. Z Zellforsch 131:347–370

Rattner JB (1972) Nuclear shaping in marsupial spermatids. J Ultrastruct Res 40:498–512

Rattner JB, Brinkley BR (1972) Ultrastructure of mammalian spermiogenesis. III. The organization and morphogenesis of the manchette during rodent spermiogenesis. J Ultrastruct Res 41:209–218

Regaud C (1901) Étude sur la structure des tubes séminifères et sur la spermatogenèse chez les mammifères. Arch Anat Microsc Morphol Exp 4:101–156

Reger JF, Fain-Maurel MA, Cassier P (1977) The origin, distribution, and fate of the chromatoid body (germ plasm) during spermatogenesis and spermiogenesis in two *Peracaridae*. J Ultrastruct Res 60:84–94

Retzius G (1909) Die Spermien der Vögel. Biologische Untersuchungen, vol XIV. Fischer, Jena, pp 89–122

Reviers M de (1968) Détermination de la durée des processus spermatogénétiques chez le coq à l'aide de thymidine tritiée. VI. Congr Int Reprod Anim Insémin Artif 1:183–184

Reviers M de (1971) Le développement testiculaire chez le coq. II. Morphologie de l'épithélium séminifère et établissement de la spermatogenèse. Ann Biol Anim Biochim Biophys 11:531–546

Reynolds ES (1963) The use of lead citrate at high pH as an electron-opaque stain in electron microscopy. J Cell Biol 17:208–212

Rhee HJ van der, Burgh-de Winter CPM van der, Daems WT (1979) The differentiation of monocytes into macrophages, epitheloid cells, and multinucleated giant cells in subcutaneous granulomas. I. Fine structure. Cell Tissue Res 197:355–378

Riley GM (1938) Cytological studies on spermatogenesis in the house sparrow, *Passer domesticus*. Cytologia (Tokyo) 9:165–176

Rolshoven E (1947/1948) Über Resorptionsleistungen des Sertoli-Syncytiums in den Hoden-Kanälchen. Anat Anz 96:220–226

Romeis B (1968) Mikroskopische Technik. Oldenbourg, München Wien

Roosen-Runge E (1955) Untersuchungen über die Degeneration samenbildender Zellen in der normalen Spermatogenese der Ratte. Z Zellforsch 41:221–235

Roosen-Runge EC (1962) The process of spermatogenesis in mammals. Biol Rev 37:343–377

Roosen-Runge EC (1973) Germinal-cell loss in normal metazoan spermatogenesis. J Reprod Fertil 35:339–348

Roosen-Runge EC (1977) The process of spermatogenesis in animals. University Press, Cambridge

Ross MH (1976) The Sertoli cell junctional specialization during spermiogenesis and at spermiation. Anat Rec 186:79–104

Rothwell B, Tingari MD (1973) The ultrastructure of the boundary tissue of the seminiferous tubule in the testis of the domestic fowl (*Gallus domesticus*). J Anat 114:321–328

Roussel JD, Stallcup OT, Austin CR (1967) Selective phagocytosis of spermatozoa in the epididymis of bulls, rabbits and monkeys. Fertil Steril 18:509–516

Rowley MJ, Berlin JD, Heller CG (1971) The ultrastructure of four types of human spermatogonia. Z Zellforsch 112:139–157

Russell L (1977a) Movement of spermatocytes from the basal to the adluminal compartment of the rat testis. Am J Anat 148:313–328

Russell L (1977b) Observations on rat Sertoli ectoplasmic ("junctional") specializations in their association with germ cells of the rat testis. Tissue Cell 9:475–498

Russell LD (1978) The blood-testis barrier and its formation relative to spermatocyte maturation in the adult rat: a Lanthanum tracer study. Anat Rec 190:99–112

Russell L, Frank B (1978) Ultrastructural characterization of nuage in spermatocytes of the rat testis. Anat Rec 190:79–98

Rutledge JT, Schwab RG (1974) Testicular metamorphosis and prolongation of spermatogenesis in starlings (*Sturnus vulgaris*) in the absence of daily photostimulation. J Exp Zool 187:71–76

Sadlier RMFS (1978) Cycles and seasons. In: Austin CR, Shorts RV (eds) Reproduction in mammals. I. Germ cells and fertilization. University Press, Cambridge, pp 85–102

Sandoz D (1970) Étude ultrastructurale et cytochimique de la formation de l'acrosome du discoglosse (amphibien anoure). In: Baccetti B (ed) Comparative spermatology. Academic Press, New York London, pp 93–113

Sapsford CS, Rae CA, Cleland KW (1969) The fate of residual bodies and degenerating germ cells and the lipid cycle in Sertoli cells in the bandicoot *Perameles nasuta* Geoffroy (*Marsupialia*). Aust J Zool 17:729–753

Schjeide OA, Nicholls T, Graham G (1972) Annulate lamellae and chromatoid bodies in the testes of a cyprinid fish (*Pimephales notatus*). Z Zellforsch 129:1–10

Schöneberg K (1913) Die Samenbildung bei den Enten. Arch Mikr Anat 83:324–369

Schulze C (1974) On the morphology of the human Sertoli cell. Cell Tissue Res 153:339–355

Schwab RG (1971) Circannian testicular periodicity in the European starling in the absence of photoperiodic change. In: Menaker M (ed) Biochronometry. Natl Acad Sci, Washington, pp 428–447

Schweigger-Seidel F (1865) Über die Samenkörperchen und ihre Entwicklung. Arch Mikr Anat 1:309–335

Serventy DL (1971) Biology of desert birds. In: Farner DS, King JR, Parkes KC (eds) Avian biology, vol I. Academic Press, New York London, pp 287–339

Skinner JD, Zyl JHM van, Heerden JAH van (1973) The effect of season on reproduction in the black wildebeest and red hartebeest in South Africa. J Reprod Fertil [Suppl] 19:101–110

Söderström KO, Parvinen M (1976) Transport of material between the nucleus, the chromatoid body and the golgi complex in the early spermatids of the rat. Cell Tissue Res 168:335–342

Stang-Voss C (1972) Ultrastrukturen der zellulären Autophagie. Elektronenmikroskopische Beobachtungen an Spermatiden von *Eisenia foetida (Annelidae)* während der cytoplasmatischen Reduktionsphase. Z Zellforsch 127:580–590

Stanley HP (1967) The fine structure of spermatozoa in the lamprey *Lampetra planeri*. J Ultrastruct Res 19:84–99

Starke FJ (1971) Elektronenmikroskopische Untersuchung der Zwittergonadenacini von *Planorbarius corneus* L. (*Basommatophora*). Z Zellforsch 119:483–514

Starke FJ, Nolte A (1970) Tubulikörper im Zytoplasma der Spermatiden von *Planorbarius corneus* L. (*Basommatophora*). Z Zellforsch 105:210–221

Susi FR, Clermont Y (1970) Fine structural modifications of the rat chromatoid body during spermiogenesis. Am J Anat 129:177–192

Swift H (1956) The fine structure of annulate lamellae. J Biophys Biochem Cytol 2:415–418

Threadgold LT (1956/1957a) The annual gonad cycle of the male jackdaw *Corvus monedulus*. Qualitative aspects. Cellule 58:19–42

Threadgold LT (1956/1957b) The annual gonad cycle of the male jackdaw *Corvus monedulus*. Quantitative aspects. Cellule 58:45–54

Tiba T, Ishikawa T, Murakami A (1968) Histologische Untersuchung der Kinetik der Spermatogenese beim Mink (*Mustela vison*). Jpn J Vet Res 16:73–85

Tingari MD (1973) Observations on the fine structure of spermatozoa in the testis and excurrent ducts of the male fowl, *Gallus domesticus*. J Reprod Fertil 34:255–265

Tingari MD, Lake PE (1972a) Ultrastructural evidence for resorption of spermatozoa and testicular fluid in the excurrent ducts of the testis of the domestic fowl, *Gallus domesticus*. J Reprod Fertil 31:373–381

Tingari MD, Lake PE (1972b) The intrinsic innervation of the reproductive tract of the male fowl (*Gallus domesticus*). A histochemical and fine structural study. J Anat 112:257–271

Tres LL, Solari AJ (1968) The ultrastructure of the nuclei and the behavior of sex chromosomes of human spermatogonia. Z Zellforsch 91:75–89

Vitale-Calpe R, Burgos MH (1970) The mechanism of spermiation in the hamster. I. Ultrastructure of spontaneous spermiation. J Ultrastruct Res 31:381–393

Wartenberg H (1978) Human testicular development and the role of the mesonephros in the origin of a dual Sertoli cell system. Andrologia 10:1–21

Wilkinson RF, Stanley HP, Bowman JT (1974) Genetic control of spermiogenesis in *Drosophila melanogaster*: the effects of abnormal cytoplasmic microtubule populations in mutant ms(3) 10 R and its colcemid-induced phenocopy. J Ultrastruct Res 48:242–258

Wing TY, Lin HS (1977) The fine structure of testicular interstitial cells in the adult golden hamster with special reference to seasonal changes. Cell Tissue Res 183:385–393

Witschi E (1935) Origin of asymmetry in the reproductive system of the birds. Am J Anat 56:119–141

Woods JE, Domm LV (1966) A histochemical identification of the androgen-producing cells in the gonads of the domestic fowl and albino rat. Gen Comp Endocrinol 7:559–570

Wright PL, Wright MH (1944) The reproductive cycle of the male red-winged blackbird. Condor 46:46–59

Yamamoto S, Tamate H, Itikawa O (1967) Morphological studies on the sexual maturation in the male Japanese quail (*Coturnix coturnix japonica*). II. The germ cell types and cellular associations during spermatogenesis. Tohoku J Agric Res 18:27–39

Yasuzumi G (1956) Electron microscopy of the developing spermhead in the sparrow testis. Exp Cell Res 11:240–243

Yasuzumi G, Sugioka T (1971) Spermatogenesis in animals as revealed by electron microscopy. XXI. Microkaryosomes and microtubules appearing during spermiogenesis of the lovebird, *Uroloncha striata var. domestica* Flower. Z Zellforsch 114:451–459

Yasuzumi G, Yamaguchi S (1977) Some aspects of the spermiogenesis in the domestic pigeon. Okajimas Folia Anat Jpn 54:139–174

Yasuzumi G, Yasuda M (1968) Spermatogenesis in animals as revealed by electron microscopy. XVIII. Fine structure of developing spermatids of the Japanese freshwater turtle fixed with potassium permanganate. Z Zellforsch 85:18–33

Zlotnik I (1947) The cytoplasmic components of germ-cells during spermatogenesis in the domestic fowl. Q J Microsc Sci 88:353–365

Subject Index

Springer-Verlag
Berlin
Heidelberg
New York

Current Topics in Microbiology and Immunology

Editors: W. Henle, P. H. Hofschneider, H. Koprowski, F. Melchers, R. Rott, H. G. Schweiger, P. K. Vogt

Volume 93
Initiation Signals in Viral Gene Expression

Editor: A. J. Shatkin

1981. 30 figures. V, 212 pages
ISBN 3-540-10804-1

Contents:
A. J. Shatkin: Introduction. – Elucidating Mechanisms of Eukaryotic Genetic Expression by Studying Animal Viruses. – *R. Tjian:* Regulation of Viral Transcription and DNA Replication by the SV40 Large T Antigen. – *T. Shenk:* Transcriptional Control Regions: Nucleotide Sequence Requirements for Initiation by RNA Polymerase II and III. – *S. J. Flint:* Splicing and the Regulation of Viral Gene Expression. – *M. Kozak:* Mechanism of mRNA Recognition by Eukaryotic Ribosomes During Initiation of Protein Synthesis. – *R. M. Krug:* Priming of Influenza Viral RNA Transcription by Capped Heterologous RNAs. – *J. Perrault:* Origin and Replication of Defective Interfering Particles. – Subject Index.

Volume 94/95

1981. 46 figures. IV, 308 pages
ISBN 3-540-10803-3

Contents:
C. W. Ward: Structure of the Influenza Virus Hemagglutinin. – *H. G. Boman, H. Steiner:* Humoral Immunity in Cecropia Pupae. – *G. Hobom:* Replication Signals in Prokaryotic DNA. – *W. Ostertag, I. B. Pragnell:* Differentiation and Viral Involvement in Differentiation of Transformed Mouse and Rat Erythroid Cells. – *J. Meyer:* Electron Microscopy of Viral RNA. – *J. Hochstadt, H. L. Ozer, C. Shopsis:* Genetic Alteration in Animal Cells in Culture.

Springer-Verlag
Berlin
Heidelberg
New York

Volume 96
Gene Cloning in Organisms other than E.coli

Editors: P. H. Hofschneider, W. Goebel

1982. 63 figures. VIII, 264 pages
ISBN 3-540-11117-4

Contents: Cloning Vectors Derived from Plasmids and Phage of Bacillus. – Use of Plasmids from Staphylococcus aureus for Cloning of DNA in Bacillus subtilis. – Vectors for Gene Cloning in Pseudomanas and Their Applications. – Host: Vector Systems for Gene Cloning in Pseudomonas. – Gene Cloning in Streptomyces. – Gene Cloning in Neurospora crassa. – Vectors for Cloning in Yeast. – Cloning with 2-μm DNA Vectors and the Expression of Foreign Genes in Saccharomyces cerevisiae. – Selectable Markers for the Transfer of Genes into Mammalian Cells. – Gene Transfer into Mammalian Cells: Use of Viral Vectors to Investigate Regulatory Signals for the Expression of Eukaryotic Genes. – Liposomes: The Development of a New Carrier System for Introducing Nucleic Acids into Plant and Animal Cells. – Cauliflower Mosaic Virus on Its Way to Becoming a Useful Plant Vector. – The Ti Plasmids of Agrobacterium. – Subject Index.

Volume 97

1982. 28 figures. IV, 204 pages
ISBN 3-540-11118-2

Contents: *M. R. Macnaughton:* The Structure and Replication of Rhinoviruses. *J. A. Holowczak:* Poxvirus DNA. *H. Persson, L. Philipson:* Regulation of Adenovirus Gene Expression. *K. H. Nierhaus:* Structure, Assembly, and Function of Ribosomes.

Volume 98
Retroviruses in Normal and Pathological Growth of Lymphocytes

Editors: E. Wecker, I. Horak

1982. 9 figures. Approx. 180 pages
ISBN 3-540-11225-1

Contents: *D. L. Steffen, H. Robinson:* Endogenous Retroviruses of Mice and Chickens. – *H. L. Robinson, G. F. Vande Woude:* The Genetic Basis of Retroviral-Induced Transformation. – *H. C. Morse III:* Expression of Xenotropic Murine Leukemia Viruses. – *E. Wecker, I. Horak:* Expression of Endogenous Viral Genes in Mouse Lymphocytes. – *E. Fleissner, H. W. Snyder, Jr.:* Oncoviral Proteins as Cellular Antigens. – *A. Schimpl:* Regulation of Lymphocyte Proliferation and Differentation by Lymphokines. – *J. N. Ihle, J. C. Lee:* Possible Immunological Mechanisms in C-Type Viral Leukemogenesis in Mice. – *I. L. Weissman, M. S. McGrath:* Retrovirus Lymphomagenesis: Relationship of Normal Immune Receptors to Malignant Cell Proliferation. – *A. Coutinho:* From the Point of View of an Immunologist. – *R. A. Weiss:* Perspectives on Endogenous Retroviruses in Normal and Pathological Growth.